Springer Tracts in Advanced Robotics

Volume 102

For further volumes:
http://www.springer.com/series/5208

STAR (Springer Tracts in Advanced Robotics) has been promoted under the auspices of EURON (European Robotics Research Network)

Giacomo Marani · Junku Yuh

Introduction to Autonomous Manipulation

Case Study with an Underwater Robot, SAUVIM

Springer

Giacomo Marani
West Virginia Robotic Technology Center
West Virginia University Research Corporation
Fairmont, WV
USA

Junku Yuh
National Agenda Research Division
Korea Institute of Science and Technology
Seoul
Korea

ISSN 1610-7438 ISSN 1610-742X (electronic)
ISBN 978-3-642-54612-9 ISBN 978-3-642-54613-6 (eBook)
DOI 10.1007/978-3-642-54613-6
Springer Heidelberg New York Dordrecht London

Library of Congress Control Number: 2014933279

Printed on acid-free paper

Springer is part of Springer Science+Business Media (www.springer.com)

Foreword

Robotics is undergoing a major transformation in scope and dimension. From a largely dominant industrial focus, robotics is rapidly expanding into human environments and vigorously engaged in its new challenges. Interacting with, assisting, serving, and exploring with humans, the emerging robots will increasingly touch people and their lives.

Beyond its impact on physical robots, the body of knowledge robotics has produced and is revealing a much wider range of applications reaching across diverse research areas and scientific disciplines, such as biomechanics, haptics, neurosciences, virtual simulation, animation, surgery, and sensor networks among others. In return, the challenges of the new emerging areas are proving an abundant source of stimulation and insights for the field of robotics. It is indeed at the intersection of disciplines that the most striking advances happen.

The *Springer Tracts in Advanced Robotics (STAR)* is devoted to bringing to the research community the latest advances in the robotics field on the basis of their significance and quality. Through a wide and timely dissemination of critical research developments in robotics, our objective with this series is to promote more exchanges and collaborations among the researchers in the community and contribute to further advancements in this rapidly growing field.

This monograph by Giacomo Marani and Junku Yuh provides an extensive tract on sensory-based autonomous manipulation for intervention tasks in unstructured environments. After presenting the theoretical foundations for kinematic and dynamic modeling as well as task-priority-based kinematic control of multibody systems, the work is focused on one of the most advanced underwater vehicle-manipulator system, SAUVIM. Solutions to the problem of target identification and localization are proposed, and a number of significant case studies are discussed.

Rich of practical examples and experimental/simulation results, this volume is the third contribution on underwater robots to the series and constitutes a fine addition to STAR!

Naples, Italy, January 2014 Bruno Siciliano

Preface

Autonomous manipulation, a challenging technology milestone, refers to the capability of a robot system that performs intervention tasks requiring physical contacts with unstructured environments without continuous human supervision. Such a robot system underlies several emerging markets and applications, including security and rescue operations, space and underwater applications, military applications, and the health-care industry.

A workshop on autonomous mobile manipulation, sponsored by NSF and NASA, was held in Houston, Texas on March 10–11, 2005, and the workshop report recognized that it is imperative to develop autonomous mobile manipulation that will instigate significant scientific, economical, and societal impact. An autonomous mobile manipulation system could be considered as an integrated system of autonomous mobility and autonomous manipulation. Robotic vehicles at a recent DARPA Grand Challenge and humanoid robots have demonstrated a certain level of autonomous mobility as technologies for mapping and navigation get matured. A certain level of dexterous manipulability has been also demonstrated in structured environments, such as industrial settings. However, autonomous manipulation based on the sensory information in unstructured environments still represents challenging research issues. This book is an introduction to this indispensable technology to the development of autonomous robots for intervention tasks.

Today's intervention tasks in general are performed with extensive human supervision, requiring high-bandwidth communication links, or in structured environments, which result in limited applications. Autonomous manipulation systems will make it possible to sense and perform mechanical work in areas that are hazardous to humans or where humans cannot go, such as natural or man-made disastrous regions, and deep-ocean, under-ice, and planetary explorations.

Autonomous manipulation systems, unlike teleoperated manipulation systems that are controlled by human operators with the aid of visual and other sensory feedback, must be capable of assessing a situation, including self-calibration based on sensory information, and executing or revising a course of manipulating action without continuous human intervention. It is sensible to consider the development of autonomous manipulation as a gradual passage from human teleoperated manipulation.

Within this passage, the most noticeable aspect is the increase of the level of information exchanged between the system and the human supervisor. In teleoperation mode, the user sends and receives low-level information in order to set the position of the manipulator with the aid of visual and other sensory feedback. As the system becomes more autonomous, the user may provide only few higher level decisional commands; the management of lower level functions (i.e., driving the motors to achieve a particular task) is left to the onboard system. The level of autonomy is related to the level of information needed by the system in performing the particular intervention. At the task execution level, the system must be capable of acting and reacting to the environment with the extensive use of sensor data processing.

This book deals with some important issues involved in the above scenario for autonomous manipulation and includes various pragmatic examples obtained during the course of a research project, SAUVIM, being jointly developed by the Autonomous Systems Laboratory (ASL) of the University of Hawaii, M.A.S.E., Inc. in Hawaii, and Naval Undersea Warfare Center (NUWC) in Rhode Island. SAUVIM represents one of the most advanced underwater robots in the world for its unique capabilities. It is a six degrees-of-freedom autonomous underwater vehicle (AUV) for 6,000 m depth, equipped with a seven degrees-of-freedom electric motor-driven robotic manipulator, eight brushless thrusters, six CPUs, and an extensive set of sensors such as imaging sonar, cameras, ultrasonic tracker, and laser ranging systems.

Chapter 2 describes modeling of mobile robots with manipulators as a multibody system. As a large body of literature exists for modeling multibody systems, this chapter presents essential elements with references, avoiding rigorous mathematics, for readers to help understanding materials in the following chapters.

Chapter 3 focuses on the manipulator control system. In autonomous systems, it must ensure a reliable behavior within the workspace, avoiding collisions, system instabilities, and unwanted motions while completing the required task, when it is theoretically executable. The control system must also address other general manipulation issues, such as task-space oriented, task priority assignment, and dynamic priority changes during the interaction with the environment.

Chapter 4 presents the full implementation of the vehicle-manipulator system. Based on the theoretical analysis introduced in the previous chapters, we introduce an innovative solution for workspace optimization in mobile manipulation, successfully implemented and tested on SAVUIM. Chapter 4 also presents our solution for optimal configuration in hovering, a critical part of an autonomous underwater intervention task. Detailed and unique experimental results for both workspace and hovering optimization are here provided.

Another important topic in autonomous manipulation is the target localization. The robot should have sensing capability to localize the target for autonomous manipulation. It can be achieved using different technologies, such as video/image processing, laser or ultrasonic 3D scanners, motion trackers (ultrasonic, magnetic, or inertial), and shape tape. Within the interaction between the robot and the environment, the automatic self-calibration is crucial. Chapter 5 describes our techniques for target localization using the DIDSON sonar (mid-range) and optical systems (short-range).

Chapter 6, as a conclusion of our work, presents the SAUVIM main software framework that allowed us to run all the presented solutions. Developing the SAUVIM software framework required a considerable effort in integrating together all the components of the system. On the top of the control system, a semi-autonomous decisional layer is proposed to perform the intervention tasks. It is supported by a communication interface (client–server architecture), a programming environment for the task execution, and a user interface environment. In this chapter we are also summarizing all the remaining technical solutions not covered by the previous ones, such as motor drivers and path planning.

Honolulu, Hawaii, March 2014

Giacomo Marani
Junku Yuh

Acknowledgments

This book would not have been possible without the help of numerous colleagues and friends.

We would like to acknowledge our gratitude for invaluable intellectual exchanges and research cooperation to the late Homayoun Seraji of NASA's Jet Propulsion Laboratory; Giuseppe Casalino of the University of Genoa, Italy; Wankyun Chung of Pohang University of Science and Technology, Korea; and Jinhyun Kim of Seoul National University of Science and Technology, Korea.

We are grateful to the U.S. Office of Naval Research for funding the SAUVIM project that produced practical examples and experimental/simulation results for this book.

We also want to thank former and current members of the ASL, M.A.S.E., and NUWC who have been involved in the SAUVIM project.

Contents

Abstract

"Autonomous manipulation" is a challenge in robotic technologies. It refers to the capability of a mobile robot system with one or more manipulators that performs intervention tasks requiring physical contacts in unstructured environments and without continuous human supervision. Achieving autonomous manipulation capability is a quantum leap in robotic technologies as it is currently beyond the state of the art in robotics.

Such robotic systems will have a real impact on our society from societal and economic points of view. This advancement will allow robots to reach new heights as they will be able to complete even more complex tasks of various applications such as security and rescue operations, space and underwater applications, military applications, flexible manufacturing, automated supply chain management, and the health care industry. Therefore, autonomous manipulation or autonomous robotic manipulation is generating an increasing interest among the entire robotics community.

Autonomous underwater manipulation is the primary research objective of the SAUVIM (Semi-Autonomous Underwater Vehicle for Intervention Missions) research group. SAUVIM (1997–2009), funded by the U.S. Office of Naval Research, was jointly developed by the Autonomous Systems Laboratory of the University of Hawaii, Marine Autonomous Systems Engineering, Inc. in Hawaii, and Naval Undersea Warfare Center Division Newport in Rhode Island.

By presenting the full research path beyond the SAUVIM program, this book addresses issues with the complexity of the problems encountered in autonomous manipulation including representation and modeling of robotic structures, kinematic and dynamic robotic control, kinematic and algorithmic singularity avoidance, dynamic task priority, workspace optimization, and environment perception. Further development in autonomous manipulation should be able to provide robust improvements of the solutions for all of the above issues.

It is hoped that this book including the findings and implications of SAUVIM would inspire the robot research community to further investigate critical issues in autonomous manipulation and to develop robot systems that can profoundly impact our society for the better.

Chapter 1
Introduction

Autonomous manipulation on a moving base, such as underwater robotic vehicles, is a very challenging task in the area of robotics, especially when accomplishing tasks in unstructured environments. This chapter presents a general introduction to the challenges of autonomous manipulation, followed by a description of the experimental platform we used in our case study: the SAUVIM autonomous underwater vehicle.

1.1 Autonomous Manipulation

Today's underwater intervention tasks are mostly performed with extensive human supervision, requiring high-bandwidth communication link, or in structured environments, which results in limited applications. Autonomous manipulation systems will make it possible to sense and perform mechanical work in areas that are hazardous to humans or where humans cannot go, such as natural or man-made disastrous regions, deep-ocean, and under-ice.

Autonomous manipulation systems, unlike teleoperated manipulation systems that are controlled by human operators with the aid of visual and other sensory feedback, must be capable of assessing a situation, including self-calibration based on sensory information, and executing or revising a course of manipulating action without continuous human intervention. It is sensible to consider the development of autonomous manipulation as a gradual passage from human teleoperated manipulation. Within this passage, the most noticeable aspect is the increase of the level of information exchanged between the system and the human supervisor.

In teleoperation with ROVs, the user sends and receives low level information in order to directly set the position of the manipulator with the aid of a visual feedback. As the system becomes more autonomous, the user may provide only a few higher level decisional commands, interacting with the task description layer. The management of lower level functions (i.e. driving the motors to achieve a particular task) is

G. Marani and J. Yuh, *Introduction to Autonomous Manipulation*,
Springer Tracts in Advanced Robotics 102, DOI: 10.1007/978-3-642-54613-6_1,

left to the onboard system. The level of autonomy is related to the level of information needed by the system in performing the particular intervention. At the task execution level, the system must be capable of acting and reacting to the environment with the extensive use of sensor data processing.

The user may provide, instead of directly operating the manipulator, higher level commands during a particular mission, such as "unplug the connector". In this approach, the function of the operator is to decide, after an analysis of the data, which particular task the vehicle is ready to execute and successively to send the decision command. The low-level control commands are provided by a pre-programmed onboard subsystem, while the virtual reality model in the local zone uses only the few symbolic information received through the low bandwidth channel in order to reproduce the actual behavior of the system.

The main approach is layered into different levels, where different behaviors take place: a low-level layer which interacts with the robot hardware, a medium-level layer for describing the control algorithms and finally a high-level layer where the task description is performed. Within this configuration, the control system for the manipulator (medium layer) must ensure a reliable behavior within the workspace, avoiding collisions, system instabilities and unwanted motions while completing the required task, when it is theoretically executable. The control system must also address other general manipulation issues, such as task-space oriented, task priority assignment, and dynamic priority changes.

AUV development is still mostly directed toward a survey-oriented vehicles. In literature there are only a few examples of Intervention AUVs. These example include the OTTER I-AUV by the Stanford Aerospace Robotics Lab. OTTER, developed back in 1996, is a hover capable underwater vehicle which operates in a test tank at the Monterey Bay Aquarium Research Institute (MBARI). Current and past research includes texture-based vision processing for feedback control and real-time mosaicking, autonomous intervention missions, and hydrodynamic modeling of underwater manipulators. A study on automatic objects retrieval was done in [1].

Another Intervention AUV, ALIVE, was developed in 2003 by Cybernetix. The aim of the EU-funded ALIVE project was to develop an Intervention-AUV capable of docking to a subsea structure which was not specifically modified for AUV use. A description of the ALIVE vehicle was given in [2].

One of the most recent research in Underwater Autonomous Manipulation is the TRIDENT project. This project proposes a new methodology to provide multipurpose dexterous manipulation capabilities for intervention operations in unknown, unstructured underwater environments. In the TRIDENT project, a multipurpose generic intervention is composed of two phases:

- PHASE I (Survey). The Autonomous Surface Craft (ASC) is launched to carry the Intervention Autonomous Underwater Vehicle (I-AUV) towards the area to be surveyed. Then, the I-AUV is deployed (1) and both vehicles start a coordinated survey path (2) to explore the area. The ASC/I-AUV team gathers navigation data for geo-referencing the measurements (seafloor images and multibeam bathymetry profiles). Finally, the I-AUV surfaces (3) and contacts to the end user to set-up and acoustic/optical map of the surveyed area. Using this map, the user selects a target object (an object of interest) as well as a suitable intervention task (grasping, hooking, etc...).

- PHASE II (Intervention). After selecting the target, the ASC/I-AUV team navigates towards the target position. Then, the ASC performs dynamic position (4) while keeping the I-AUV inside the USBL cone of coverage. Then, the I-AUV performs a search (5) looking for the Target of Interest (ToI). When the object appears in the robot field of view, it is identified and the I-AUV switches to free floating mode using its robotic arm as well as the dexterous hand to do the manipulation (6). Finally (7), the I-AUV docks to the ASC before recovery.

In this book, we present the latest achievements with the SAUVIM AUV. SAUVIM (Fig. 1.1) was jointly developed by the Autonomous Systems Laboratory (ASL) of the University of Hawaii, Marine Autonomous Systems Engineering (MASE), Inc. in Hawaii, and Naval Undersea Warfare Center Division Newport (NUWC) in Rhode Island.

1.2 State of the Art: SAUVIM for Autonomous Intervention Missions

SAUVIM (Semi Autonomous Underwater Vehicle for Intervention Missions, [3–5], Fig. 1.1) involves the design and fabrication of an underwater vehicle that it is capable of autonomous interventions on the subsea installations, a task usually carried out by ROVs or human divers. The vehicle is built around an open-framed structure enclosed by a flooded composite fairing. Its movement is controlled by eight thrusters located around the center of mass. The four vertical thrusters move the vehicle in the Z-axis (heave); the two, internally mounted, horizontal thrusters move the vehicle in the Y-axis (sway); and the two, externally-mounted, horizontal thrusters move the vehicle in the X-axis (surge). The lower frame houses only the NI-MH battery pack, while the upper frame hosts all the essential electronics, visual hardware, navigation and mission sensors in six cylindrical pressure vessels.

The proper subsea navigation and positioning accuracy is achieved with a Photonic Inertial Navigation System (PHINS) unit from IXSEA. This INS outputs position, heading, roll, pitch, depth, velocity, and heave. Its high accuracy inertial measurement capability is based on Fiber Optic Gyroscope technology coupled with an embedded digital signal processor that runs an advanced Kalman filter. This INS is aided by a differential GPS, a Doppler Velocity Log (DVL) other than the classical depth sensor, in order to improve the absolute measurement of the vehicle position.

This navigation sensor system is capable of providing a stable and precise feedback of the vehicle position, velocity and acceleration on all the 6 DOF. In one of our experiments, we tested the stability and precision of the vehicle during station keeping with a generalized PID controller active on all 6 axis. The INS was aided only by the DVL, since the GPS was, in this case, submerged. The vehicle was able to maintain the

Fig. 1.1 The SAUVIM autonomous underwater vehicle

target position with a sub-inch accuracy for the translational part. This was confirmed by the manipulator camera output, which was looking toward an earth-fixed target. This position was maintained for over 15 m without any relevant change in the x and y axis positions. We only noticed a slow change in the Z position, which was due to the tide activity. As a matter of fact, while the target was fixed with respect to the earth, the INS uses a depth sensor to correct the z-coordinate. A video recording of the above experiment can be downloaded from the Autonomous Systems Laboratory web sites.[1]

To achieve the intervention capabilities, SAUVIM is equipped with a 7 degrees-of-freedom robotic manipulator (MARIS 7080, Fig. 1.2). The arm, unlike the classical hydraulic arm in use for ROVs, is actuated by electromechanical components, in order to meet the low-power requirements and higher accuracy in manipulation tasks. Each degree of freedom is actuated by a brushless motor with a reduction unit (harmonic drive). The accuracy of the angular measurement is guaranteed by the combination of two resolvers, mounted respectively before and after the reduction unit. This configuration allows a sub-millimeter positioning of the manipulator's end-effector, an important requirement when dealing with a large class of underwater intervention. A force/torque sensor, installed between the last degree of freedom of the wrist and the gripper, senses the amount of the force and torque acting on the gripper. Designed for underwater applications at extreme depths, it is internally compensated with appropriate oil.

The sensor devices of SAUVIM are the most critical components of a generic intervention mission, since at the task execution level the system must be capable of acting and reacting to the environment with the extensive use of sensor data

[1] http://gmarani.org/sauvim

Fig. 1.2 MARIS 7080 underwater manipulator

processing. SAUVIM is equipped with a Dual Frequency Identification Sonar (DIDSON[2]), a Digital Multi-Frequency Scanning Sonar, several video camera with Image Processing computational unit and a special ultrasonic device for tracking the position of a generic target in 6 degrees-of-freedom.

Finally, the hardware architecture is composed of several computers and peripherals for sensor data acquisition. Two VME-based single board computers host the distributed control system for the vehicle and manipulator, and 3 PC-104 units provide the necessary computation power for sensor data processing, including a dedicated system for optical vision.

References

1. Wang HH, Rock SM, Lee MJ (1995) Experiments in the automatic retrieval of underwater objects with an AUV. In: Proceedings of oceans 95, MTS/IEEE, MTS/IEEE, San Diego, pp 366–373. http://www.stanford.edu/group/arl/cgi-bin/drupal/sites/default/files/public/publications/WangR95.pdf

[2] http://www.soundmetrics.com

2. Evans J, Redmond P, Plakas C, Hamilton K, Lane D (2003) Autonomous docking for intervention-auvs using sonar and video-based real-time 3d pose estimation. In: Oceans 2003. Proceedings, vol 4, pp 2201–2210. doi:10.1109/OCEANS.2003.178243
3. Marani G, Choi SK, Yuh J (2009) Underwater autonomous manipulation for intervention missions auvs. Ocean Eng 36(1):15–23. doi:10.1016/j.oceaneng.2008.08.007, http://www.sciencedirect.com/science/article/B6V4F-4T7F5PD-1/2/ce310fd84710cc86d5ffd871b95edd70,aUV
4. Yuh J, Choi S, Ikehara C, Kim G, McMurty G, Ghasemi-Nejhad M, Sarkar N, Sugihara K (1998) Design of a semi-autonomous underwater vehicle for intervention missions (sauvim). In: Underwater technology, 1998. Proceedings of the 1998 international symposium on, pp 63–68. doi:10.1109/UT.1998.670059
5. Yuh J, Choi SK (1999) Semi-autonomous underwater vehicle for intervention mission: An auv that does more than just swim. Sea Technol 40(10):37–42

Chapter 2
Geometry, Kinematics and Dynamics of Multi-body Systems

The search for a mathematical model to accurately represent the physical behavior of a generic mechanical system is the most important step in the development of simulation, identification and control.

With the constant growth of the capabilities of modern computing systems, the limit of the complexity of a mathematical model can be increased to describe a larger category of structures, yet maintaining the possibility of performing simulation, identification and/or control in real time.

Moreover, the availability of powerful tools for symbolic computation enables the possibility of automating the entire process that, starting from a structure description, performs all elaborations necessary to compute the model. This concept requires the introduction of models built in a way that is independent on the particular structure to which they refers.

This approach is particularly useful when dealing with mobile manipulation. Conventionally, mobile manipulation systems have been often regarded as composition of two or more (possibly coupled) structures: the robotic arm(s) and the vehicle. By introducing the generalization of a mechanical joint with variable degrees of freedom, it is possible to regard the entire system as an unique "robot", in general with a multi-branch topology, and made of different types of joints. Within this approach concepts like manipulability can easily be applied to the whole vehicle-manipulator system and, as introduced later in the book, can be particularly useful in solving important problems in autonomous manipulation like the workspace optimization.

This chapter serves as an introduction to general notions and concepts on geometry, kinematics and dynamics of multi-body system. As a large body of literature exists for modeling multi-body systems, this chapter presents only the essential elements to understand our approach in modeling the mobile vehicle-manipulation system as a generalized multi-body structure.

G. Marani and J. Yuh, *Introduction to Autonomous Manipulation*,
Springer Tracts in Advanced Robotics 102, DOI: 10.1007/978-3-642-54613-6_2,

2.1 Geometry

This section focuses on basic geometrical concepts applied to multi-body systems. We will introduce the concepts of reference system transformation, to prepare a background necessary to describe a generic robotic structure.

2.1.1 Vector Operations

In this section we assume that the reader is familiar with basic analytic geometry concepts, such as free and oriented vectors.

2.1.1.1 Scalar Product

Consider two generic free geometric vectors $\boldsymbol{v}_1$ and $\boldsymbol{v}_2$. The *scalar product* (*dot product*) is represented and defined as:

$$\boldsymbol{v_1} \cdot \boldsymbol{v_2} = |\boldsymbol{v_1}|\,|\boldsymbol{v}_2| \cos(\theta) \tag{2.1}$$

with θ the minimum angle between the two vectors. Assuming $\boldsymbol{v} = \begin{bmatrix} v_1 \; v_2 \; v_3 \end{bmatrix}^T$ and $\boldsymbol{w} = \begin{bmatrix} w_1 \; w_2 \; w_3 \end{bmatrix}^T$. Equation (2.1) results in the following matrix form:

$$\boldsymbol{v} \cdot \boldsymbol{w} = \boldsymbol{v}^T \boldsymbol{w} = \begin{bmatrix} v_1 \; v_2 \; v_3 \end{bmatrix} \begin{bmatrix} w_1 \\ w_2 \\ w_3 \end{bmatrix} = v_1\,w_1 + v_2\,w_2 + v_3\,w_3 \tag{2.2}$$

The dot product is commutative:

$$\boldsymbol{v}_1 \cdot \boldsymbol{v}_2 = \boldsymbol{v}_2 \cdot \boldsymbol{v}_1 \tag{2.3}$$

and distributive with respect to the addition:

$$\boldsymbol{w} \cdot (\boldsymbol{v}_1 + \boldsymbol{v}_2) = \boldsymbol{w} \cdot \boldsymbol{v}_1 + \boldsymbol{w} \cdot \boldsymbol{v}_2. \tag{2.4}$$

2.1.1.2 Cross Product

Given two generic free vectors $\boldsymbol{v}_1$ and $\boldsymbol{v}_2$, the *cross product* is defined as (Fig. 2.1):

$$\boldsymbol{v_1} \times \boldsymbol{v_2} = |\boldsymbol{v_1}|\,|\boldsymbol{v}_2| \sin(\theta)\,\boldsymbol{n} \tag{2.5}$$

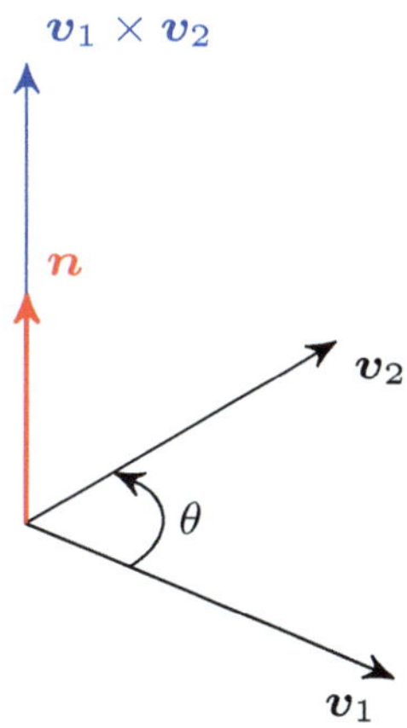

Fig. 2.1 Cross product

where θ is the the measure of the smaller angle between $\boldsymbol{v}_1$ and $\boldsymbol{v}_2$ and $\boldsymbol{n}$ is an unit vector perpendicular to the plane containing $\boldsymbol{v}_1$ and $\boldsymbol{v}_2$, with the direction given by the right-hand rule. Important properties of the cross product are the following:

- $\boldsymbol{v}_1 \times \boldsymbol{v}_2 = -\boldsymbol{v}_2 \times \boldsymbol{v}_1$ (anticommutative)
- $\boldsymbol{v}_1 \times (\boldsymbol{v}_2 + \boldsymbol{v}_3) = \boldsymbol{v}_1 \times \boldsymbol{v}_2 + \boldsymbol{v}_1 \times \boldsymbol{v}_3$ (distributive over addition).

2.1.2 Reference Systems

In robotics, besides an absolute Cartesian coordinate system (*main frame*), it is necessary to distribute local reference frames, integral on each link of the manipulator. The position of each frame with respect to the main frame must be described by specifying the origin and direction of the coordinate axes (Fig. 2.2):

$$\langle a \rangle \equiv \{O_a, \boldsymbol{i}_a, \boldsymbol{j}_a, \boldsymbol{k}_a\}$$
$$\langle b \rangle \equiv \{O_b, \boldsymbol{i}_b, \boldsymbol{j}_b, \boldsymbol{k}_b\}$$

Any vector can be uniquely expressed w.r.t. either $\langle a \rangle$ or $\langle b \rangle$ frame:

$$\boldsymbol{v} = c_1 \boldsymbol{i}_a + c_2 \boldsymbol{j}_a + c_3 \boldsymbol{k}_a \tag{2.6}$$
$$\boldsymbol{v} = h_1 \boldsymbol{i}_a + h_2 \boldsymbol{j}_a + h_3 \boldsymbol{k}_a. \tag{2.7}$$

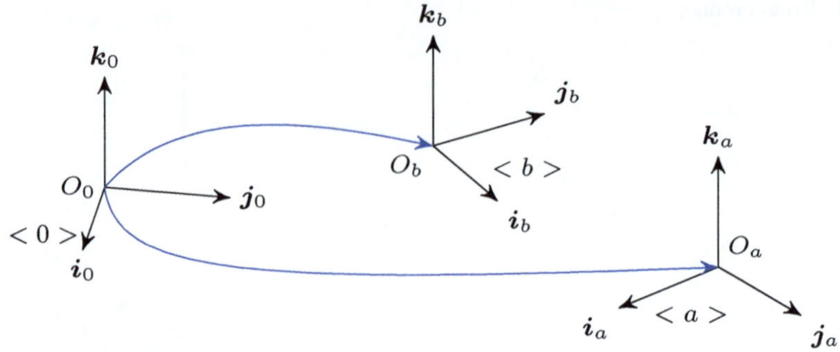

Fig. 2.2 Reference frames

2.1.2.1 Orthogonal Reference Frames

The frame $\langle a \rangle \equiv \{O_a, \boldsymbol{i}_a, \boldsymbol{j}_a, \boldsymbol{k}_a\}$ is *right-handed* if:

$$\begin{cases} \boldsymbol{k}_a = \boldsymbol{i}_a \times \boldsymbol{j}_a \\ \boldsymbol{i}_a \cdot \boldsymbol{j}_a = 0 \end{cases} \tag{2.8}$$

Similarly, the frame $\langle a \rangle \equiv \{O_a, \boldsymbol{i}_a, \boldsymbol{j}_a, \boldsymbol{k}_a\}$ is *left-handed* if:

$$\begin{cases} \boldsymbol{k}_a = -\boldsymbol{i}_a \times \boldsymbol{j}_a \\ \boldsymbol{i}_a \cdot \boldsymbol{j}_a = 0 \end{cases}. \tag{2.9}$$

2.1.2.2 Algebraic Vectors

The concept of geometric vector can be defined algebraically by placing the vector in a rectangular coordinate system. The components of $\boldsymbol{v}$ w.r.t. a reference frame can be organized in a 3×1 matrix known as *algebraic vector*. Thus, in case of Eq. (2.6), the equivalent algebraic vector is given by:

$${}^{a}\boldsymbol{v} = \begin{bmatrix} c_1 \\ c_2 \\ c_3 \end{bmatrix} \in \Re^3 \tag{2.10}$$

Vector ${}^{a}\boldsymbol{v}$ is the projection of $\boldsymbol{v}$ on the frame $\langle a \rangle$. Similarly we have:

$${}^{b}\boldsymbol{v} = \begin{bmatrix} h_1 \\ h_2 \\ h_3 \end{bmatrix} \in \Re^3 \tag{2.11}$$

where ${}^{b}\boldsymbol{v}$ is the projection of $\boldsymbol{v}$ on the frame $\langle b \rangle$.

The vector operations introduced in Sect. 2.1.1 apply also to the algebraic vectors, in the same reference frame. In case the reference frame is orthonormal we have:

- dot product:

$$\boldsymbol{v} \cdot \boldsymbol{w} = {}^{a}\boldsymbol{v}^{T}\, {}^{a}\boldsymbol{w} \tag{2.12}$$

- cross product in right-handed frame:

$${}^{a}(\boldsymbol{v} \times \boldsymbol{w}) = \lfloor {}^{a}\boldsymbol{v} \times \rfloor\, {}^{a}\boldsymbol{w} \tag{2.13}$$

- cross product in left-handed frame:

$${}^{a}(\boldsymbol{v} \times \boldsymbol{w}) = -\lfloor {}^{a}\boldsymbol{v} \times \rfloor\, {}^{a}\boldsymbol{w} \tag{2.14}$$

where $\lfloor {}^{a}\boldsymbol{v} \times \rfloor$ is the the skew-symmetric matrix operator, defined as (considering Eq. (2.6)):

$$\lfloor {}^{a}\boldsymbol{v} \times \rfloor = \begin{bmatrix} 0 & -c_3 & c_2 \\ c_3 & 0 & -c_1 \\ -c_2 & c_1 & 0 \end{bmatrix}. \tag{2.15}$$

2.1.2.3 Rotation Matrix

The relative orientation of two reference frames can be described with the concept of rotation matrix. Given two reference systems $\langle a \rangle \equiv \{O_a, \boldsymbol{i}_a, \boldsymbol{j}_a, \boldsymbol{k}_a\}$ and $\langle b \rangle \equiv \{O_b, \boldsymbol{i}_b, \boldsymbol{j}_b, \boldsymbol{k}_b\}$, the *rotation matrix* of the frame $\langle b \rangle$ w.r.t. the frame $\langle a \rangle$ is the 3×3 matrix defined as:

$${}^{a}_{b}\boldsymbol{R} \hat{=} \left[{}^{a}\boldsymbol{i}_b \; {}^{a}\boldsymbol{j}_b \; {}^{a}\boldsymbol{k}_b \right] \in \Re^{3\times 3} \tag{2.16}$$

Assuming that both $\langle a \rangle$ and $\langle b \rangle$ are orthonormal, the rotation matrix has the following properties:

$$\begin{cases} {}^{a}\boldsymbol{i}_b \cdot {}^{a}\boldsymbol{j}_b = 0 \\ {}^{a}\boldsymbol{j}_b \cdot {}^{a}\boldsymbol{k}_b = 0 \\ {}^{a}\boldsymbol{k}_b \cdot {}^{a}\boldsymbol{i}_b = 0 \end{cases} \tag{2.17}$$

$$\begin{cases} {}^{a}\boldsymbol{i}_b \cdot {}^{a}\boldsymbol{i}_b = 1 \\ {}^{a}\boldsymbol{j}_b \cdot {}^{a}\boldsymbol{j}_b = 1 \\ {}^{a}\boldsymbol{k}_b \cdot {}^{a}\boldsymbol{k}_b = 1 \end{cases} \tag{2.18}$$

$${}^{a}_{b}\boldsymbol{R}^{-1} = {}^{a}_{b}\boldsymbol{R}^{T} \tag{2.19}$$

$$\det\left({}^a_b\boldsymbol{R}\right) = \pm 1 \tag{2.20}$$

Note that the sign in Eq. (2.20) is positive in case both $\langle a \rangle$ and $\langle b \rangle$ frames are right-handed or left-handed, and negative if they are opposite (one left-handed and the other right-handed).

2.1.2.4 Change of Reference System

Given a geometric free vector $\boldsymbol{v}$, whose projection ${}^b\boldsymbol{v}$ on the frame $\langle b \rangle$ is known, the problem in changing reference system consist in finding the projection ${}^a\boldsymbol{v}$ of the same vector on the frame $\langle a \rangle$. Considering the geometric vector

$$\boldsymbol{v} = x_b\boldsymbol{i}_b + y_b\boldsymbol{j}_b + z_b\boldsymbol{k}_b \tag{2.21}$$

its projection on the frame $\langle a \rangle$ is:

$${}^a\boldsymbol{v} = x_b{}^a\boldsymbol{i}_b + y_b{}^a\boldsymbol{j}_b + z_b{}^a\boldsymbol{k}_b \tag{2.22}$$

Considering Eq. (2.16), the above projection becomes:

$${}^a\boldsymbol{v} = {}^a_b\boldsymbol{R}\,{}^b\boldsymbol{v} \tag{2.23}$$

Similarly we have:

$${}^b\boldsymbol{v} = {}^b_a\boldsymbol{R}\,{}^a\boldsymbol{v} \tag{2.24}$$

that, considering Eqs. (2.19) and (2.23), leads to:

$${}^b_a\boldsymbol{R} = {}^a_b\boldsymbol{R}^T \tag{2.25}$$

This means that, while the columns of the rotation matrix ${}^a_b\boldsymbol{R}$ are the unit vector axes of $\langle b \rangle$ projected onto $\langle a \rangle$, its rows are the unit vector axes of $\langle a \rangle$ projected onto $\langle b \rangle$.

2.1.2.5 Change of Reference System for the Skew Symmetric Operator

Let's consider now how the skew symmetric matrix operator introduced in Eq. (2.15) changes with the reference frame. Given two vectors $\boldsymbol{v}$ and $\boldsymbol{w}$, their cross product projected on the frame $\langle a \rangle$ is given by:

$${}^a(\boldsymbol{v} \times \boldsymbol{w}) = \left\lfloor {}^a\boldsymbol{v} \times \right\rfloor {}^a\boldsymbol{w} \tag{2.26}$$

Similarly, the cross product projected on a different frame $\langle b \rangle$ is given by:

$$ {}^{b}(\boldsymbol{v} \times \boldsymbol{w}) = \left\lfloor {}^{b}\boldsymbol{v} \times \right\rfloor {}^{b}\boldsymbol{w} \tag{2.27} $$

Projecting the left hand side of Eq. (2.27) on the frame $\langle a \rangle$, and considering Eq. (2.26) we obtain:

$$ \begin{aligned} {}^{a}\left({}^{b}(\boldsymbol{v} \times \boldsymbol{w})\right) &= {}^{a}_{b}\boldsymbol{R}\, {}^{b}(\boldsymbol{v} \times \boldsymbol{w}) \\ &= {}^{a}_{b}\boldsymbol{R} \left\lfloor {}^{b}\boldsymbol{v} \times \right\rfloor {}^{b}\boldsymbol{w} \\ &= {}^{a}_{b}\boldsymbol{R} \left\lfloor {}^{b}\boldsymbol{v} \times \right\rfloor {}^{a}_{b}\boldsymbol{R}^{T}\, {}^{a}\boldsymbol{w} \end{aligned} \tag{2.28} $$

from which, for the arbitrariness of $\boldsymbol{w}$, we obtain:

$$ \left\lfloor {}^{a}\boldsymbol{v} \times \right\rfloor = {}^{a}_{b}\boldsymbol{R} \left\lfloor {}^{b}\boldsymbol{v} \times \right\rfloor {}^{a}_{b}\boldsymbol{R}^{T}. \tag{2.29} $$

2.1.2.6 Concatenation of Rotation Matrices

Given a certain number of reference frames (see Fig. 2.3), let's now determine the rotation matrix ${}^{h}_{k}\boldsymbol{R}$ of the generic frame $\langle k \rangle$ w.r.t. the frame $\langle h \rangle$, i.e. the matrix such as:

$$ {}^{h}\boldsymbol{v} = {}^{h}_{k}\boldsymbol{R}\, {}^{k}\boldsymbol{v} \tag{2.30} $$

At this aim let's consider a path, starting on frame $\langle k \rangle$ and possibly without loops, that connects the two frames. Along the path, let's project the vector ${}^{k}\boldsymbol{v}$ on each successive frame, by pre-multiplying by each rotation matrix. With reference to the path k-h in Fig. 2.3, we obtain:

$$ {}^{h}\boldsymbol{v} = {}^{h}_{j}\boldsymbol{R}\, {}^{j}_{i}\boldsymbol{R}\, {}^{i}_{k}\boldsymbol{R}\, {}^{k}\boldsymbol{v} \tag{2.31} $$

from which:

$$ {}^{h}_{k}\boldsymbol{R} = {}^{h}_{j}\boldsymbol{R}\, {}^{j}_{i}\boldsymbol{R}\, {}^{i}_{k}\boldsymbol{R}. \tag{2.32} $$

2.1.2.7 Representation of the Rotation Matrix

Because the 6 constraints introduced in Eqs. (2.17) and (2.18), the nine-elements rotation matrix can be represented with a minimum of 3 parameters. In literature there are several methodologies to represent a rotation matrix; here, we are introducing only the representations used on our work.

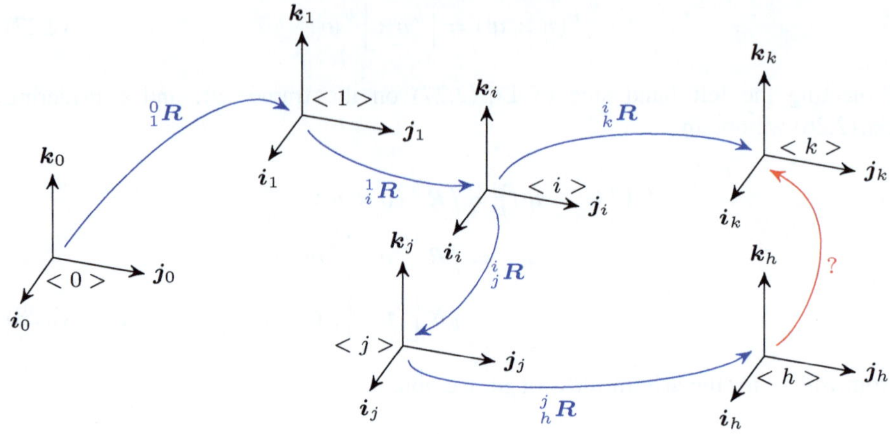

Fig. 2.3 Reference frames

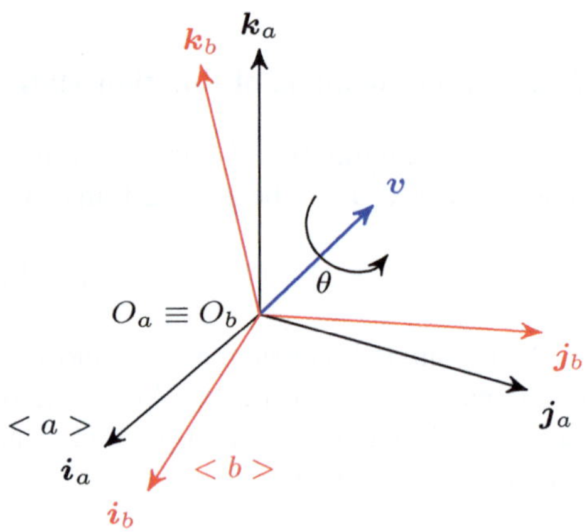

Fig. 2.4 Frame rotation: frame $\langle b \rangle$ is obtained by rotating frame $\langle a \rangle$ of 15° around vector $^a\boldsymbol{v} = [0.54\ 0.54\ 0.65]^T$

Exponential representation

The frame $\langle b \rangle$, initially coincident with $\langle a \rangle$, is being rotated of an angle θ around the unity vector $\boldsymbol{v}$ passing through the origin (Fig. 2.4). The rotation matrix $^a_b\boldsymbol{R}$ of the frame $\langle b \rangle$ w.r.t. $\langle a \rangle$ can be expressed as:

$$^a_b\boldsymbol{R} = \boldsymbol{R}(\boldsymbol{v}, \theta) = e^{\lfloor \boldsymbol{v} \times \rfloor \theta} \tag{2.33}$$

with $\lfloor \boldsymbol{v} \times \rfloor$ defined as in Eq. (2.15). Expanding the matrix exponential we obtain:

$$e^{\lfloor \boldsymbol{v} \times \rfloor \theta} = \begin{bmatrix} v_x v_x p\theta + c\theta & v_x v_y p\theta - v_z s\theta & v_x v_z p\theta + v_y s\theta \\ v_x v_y p\theta + v_z s\theta & v_y v_y p\theta + c\theta & v_y v_z p\theta - v_x s\theta \\ v_x v_z p\theta - v_y s\theta & v_y v_z p\theta + v_x s\theta & v_z v_z p\theta + c\theta \end{bmatrix} \tag{2.34}$$

where $c\theta \hat{=} \cos(\theta)$, $s\theta \hat{=} \sin(\theta)$ and $p\theta \hat{=} 1-\cos(\theta)$. The exponential representation (2.33) has the following properties:

1. Symmetry:

$$\boldsymbol{R}(\boldsymbol{v},\theta) = \boldsymbol{R}(-\boldsymbol{v},\theta) \tag{2.35}$$

2. Periodicity:

$$\boldsymbol{R}(\boldsymbol{v},\theta) = \boldsymbol{R}(\boldsymbol{v},\theta \pm 2k\pi) \tag{2.36}$$

3. Identity:

$$\boldsymbol{R}(\boldsymbol{v},2k\pi) = \boldsymbol{I} \quad \forall \boldsymbol{v} \tag{2.37}$$

The rotation matrix is thus fully determined by the unity vector $\boldsymbol{v}$ and the angle θ (see Fig. 2.4). Proofs of Eqs. (2.33) and (2.34) are out of the scope of this book and hence omitted.

Examples

- $\boldsymbol{v} = \boldsymbol{i}$

$$\boldsymbol{R}(\boldsymbol{i},\theta) = \begin{bmatrix} 1 & 0 & 0 \\ 0 & \cos(\theta) & -\sin(\theta) \\ 0 & \sin(\theta) & \cos(\theta) \end{bmatrix} \tag{2.38}$$

- $\boldsymbol{v} = \boldsymbol{j}$

$$\boldsymbol{R}(\boldsymbol{j},\theta) = \begin{bmatrix} \cos(\theta) & 0 & \sin(\theta) \\ 0 & 1 & 0 \\ -\sin(\theta) & 0 & \cos(\theta) \end{bmatrix} \tag{2.39}$$

- $\boldsymbol{v} = \boldsymbol{k}$

$$\boldsymbol{R}(\boldsymbol{k},\theta) = \begin{bmatrix} \cos(\theta) & -\sin(\theta) & 0 \\ \sin(\theta) & \cos(\theta) & 0 \\ 0 & 0 & 1 \end{bmatrix} \tag{2.40}$$

Euler roll-pitch-yaw representation

Let's consider two frames $\langle a \rangle$ and $\langle b \rangle$ and two auxiliary frames $\langle b_1 \rangle$ and $\langle b_2 \rangle$, all initially coincident with $\langle a \rangle$ (see Fig. 2.5). The final placement of $\langle b \rangle$ w.r.t. $\langle a \rangle$ is obtained through the following steps:

1. Rotate the frame $\langle b_1 \rangle$ w.r.t. $\langle a \rangle$ of the angle α (*yaw*) around its own axis $\boldsymbol{k}_1$. Frame $\langle b_1 \rangle$, during its rotation, carries the frames $\langle b_2 \rangle$ and $\langle b \rangle$.
2. Rotate the frame $\langle b_2 \rangle$ w.r.t. $\langle b_1 \rangle$ of the angle β (*pitch*) around its own axis $\boldsymbol{j}_2$. Frame $\langle b_2 \rangle$, during its rotation, carries the frame $\langle b \rangle$.
3. Rotate the frame $\langle b \rangle$ w.r.t. $\langle b_2 \rangle$ of the angle γ (*roll*) around its own axis $\boldsymbol{i}_3$.

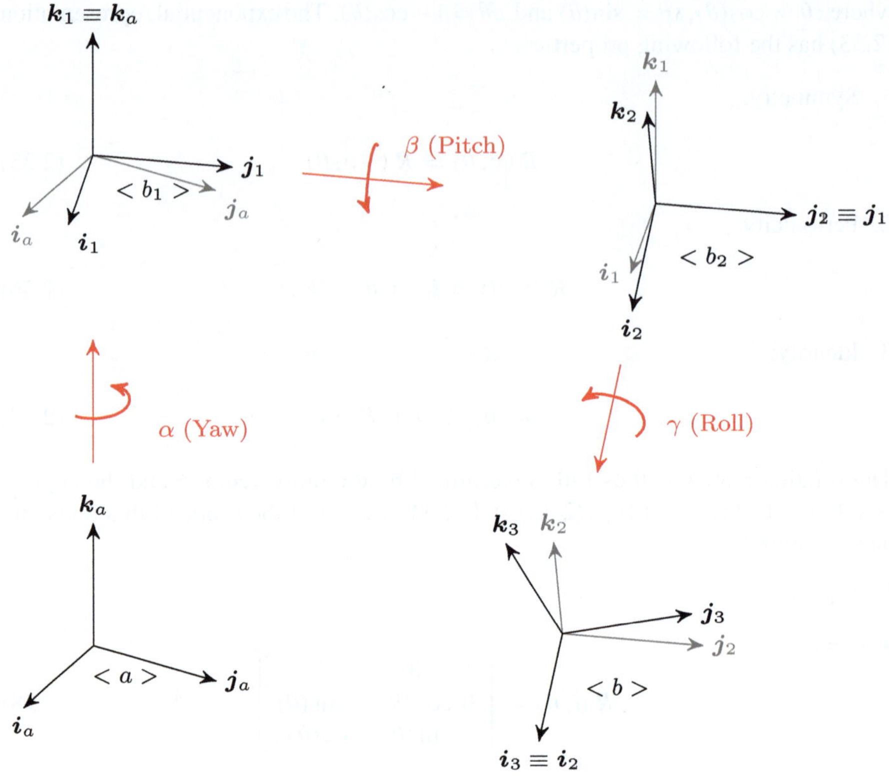

Fig. 2.5 Euler angles: case of $\alpha = 20°$, $\beta = 20°$ and $\gamma = 20°$

The final rotation matrix becomes:

$$
\begin{aligned}
{}^{a}_{b}\boldsymbol{R} &= \boldsymbol{R}(\boldsymbol{k}, \alpha)\boldsymbol{R}(\boldsymbol{j}, \beta)\boldsymbol{R}(\boldsymbol{i}, \gamma) \\
&= e^{\lfloor \boldsymbol{k}\times \rfloor \alpha} e^{\lfloor \boldsymbol{j}\times \rfloor \beta} e^{\lfloor \boldsymbol{i}\times \rfloor \gamma}
\end{aligned}
\tag{2.41}
$$

Expanding Eq. (2.41) we obtain:

$$
\begin{bmatrix}
\cos(\beta)\cos(\alpha) & -\cos(\gamma)\sin(\alpha) + \sin(\gamma)\sin(\beta)\cos(\alpha) & \sin(\gamma)\sin(\alpha) + \cos(\gamma)\sin(\beta)\cos(\alpha) \\
\cos(\beta)\sin(\alpha) & \cos(\gamma)\cos(\alpha) + \sin(\gamma)\sin(\beta)\sin(\alpha) & -\sin(\gamma)\cos(\alpha) + \cos(\gamma)\sin(\beta)\sin(\alpha) \\
-\sin(\beta) & \sin(\gamma)\cos(\beta) & \cos(\gamma)\cos(\beta)
\end{bmatrix}
\tag{2.42}
$$

The inverse problem, i.e. finding the Euler angles α, β and γ from a rotation matrix, is not unique. We utilize the following algorithm:

1. γ (roll).
   Note that:

$$\boldsymbol{i}_2 = \boldsymbol{i}_3 \tag{2.43}$$

which implies that $\boldsymbol{j}_2$ is orthogonal to $\boldsymbol{i}_3$. Similarly, $\boldsymbol{j}_2 = \boldsymbol{j}_1$ implies that $\boldsymbol{j}_2$ is orthogonal to $\boldsymbol{k}_0$. Hence, if $\boldsymbol{k}_0$ and $\boldsymbol{i}_3$ are not parallel, we have:

$$\boldsymbol{j}_2 = \pm\frac{\boldsymbol{k}_0 \times \boldsymbol{i}_3}{|\boldsymbol{k}_0 \times \boldsymbol{i}_3|} \tag{2.44}$$

$\boldsymbol{k}_2$ can then be determined from Eqs. (2.44) and (2.43):

$$\boldsymbol{k}_2 = \boldsymbol{i}_2 \times \boldsymbol{j}_2 \tag{2.45}$$

The roll angle γ can be now easily computed from the relationships:

$$\cos(\gamma) = \boldsymbol{j}_3 \cdot \boldsymbol{j}_2 \tag{2.46}$$

and

$$\sin(\gamma) = \boldsymbol{j}_3 \cdot \boldsymbol{k}_2 \tag{2.47}$$

obtaining:

$$\gamma = \arg\left[(\boldsymbol{j}_3 \cdot \boldsymbol{k}_2), (\boldsymbol{j}_3 \cdot \boldsymbol{j}_2)\right] \tag{2.48}$$

2. β (pitch)
   Considering that the axes of the frame $\langle b_1 \rangle$ have the following properties:

$$\boldsymbol{j}_1 = \boldsymbol{j}_2$$

$$\boldsymbol{k}_1 = \boldsymbol{k}_0$$

$$\boldsymbol{i}_1 = \boldsymbol{j}_1 \times \boldsymbol{k}_1 = \boldsymbol{j}_2 \times \boldsymbol{k}_0$$

we can determine the pitch angle as following:

$$\cos(\beta) = \boldsymbol{k}_2 \cdot \boldsymbol{k}_1$$

$$\sin(\beta) = \boldsymbol{k}_2 \cdot \boldsymbol{i}_1$$

and finally:

$$\beta = \arg\left[(\boldsymbol{k}_2 \cdot \boldsymbol{i}_1), (\boldsymbol{k}_2 \cdot \boldsymbol{k}_1)\right] \tag{2.49}$$

3. α (yaw)
   Since:

$$\cos(\alpha) = \boldsymbol{i}_1 \cdot \boldsymbol{i}_0$$

$$\sin(\alpha) = \boldsymbol{i}_1 \cdot \boldsymbol{j}_0$$

we simply have:

$$\alpha = \arg\left[\left(\boldsymbol{i}_1 \cdot \boldsymbol{j}_0\right), \left(\boldsymbol{i}_1 \cdot \boldsymbol{i}_0\right)\right] \tag{2.50}$$

The angles obtained with the positive sign in Eq. (2.44) are the ones shown in Fig. 2.5. In the above procedure we have assumed that $\boldsymbol{k}_0$ and $\boldsymbol{i}_3$ are not parallel. In case they are parallel (easy to verify), the procedure changes by noting that:

$$\beta = \pm\frac{\pi}{2} \tag{2.51}$$

For the remaining angles, there is an infinity of solutions that satisfy the relationship:

$$\alpha + \gamma = \xi \tag{2.52}$$

where ξ is a constant. The latter is obtained by considering that:

$$\cos(\xi) = \boldsymbol{k}_3 \cdot \boldsymbol{i}_0$$

$$\sin(\xi) = \boldsymbol{k}_3 \cdot \boldsymbol{j}_0$$

hence:

$$\xi = \arg\left[\left(\boldsymbol{k}_3 \cdot \boldsymbol{j}_0\right), \left(\boldsymbol{k}_3 \cdot \boldsymbol{i}_0\right)\right]. \tag{2.53}$$

2.1.2.8 The Transformation Matrix

After considering the transformation of free geometric vectors, let's now introduce the change of reference system for points in the Euclidian space.

Given a point P, its projection on the frames $\langle a \rangle$ and $\langle b \rangle$, visualized in Fig. 2.6, is:

$${}^{a}P = {}^{a}(P - O_a)$$

$${}^{b}P = {}^{b}(P - O_b)$$

In general, unlike the free geometric vectors, we have:

$${}^{a}\left({}^{b}P\right) \neq {}^{a}P \tag{2.54}$$

and also:

$${}^{a}\left({}^{b}P\right) = {}^{a}\left({}^{b}(P - O_b)\right) = {}^{a}(P - O_b) \tag{2.55}$$

The relationship between ${}^{a}P$ and ${}^{b}P$ is:

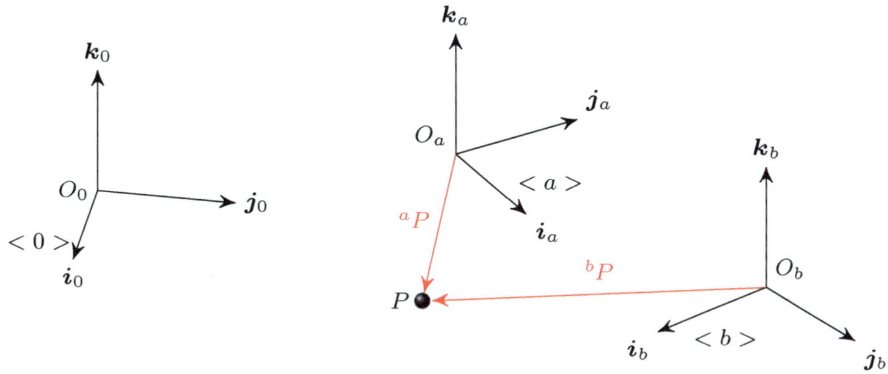

Fig. 2.6 Point transformation

$$^aP = {^aO_b} + {^b_a\boldsymbol{R}}{^bP} \tag{2.56}$$

Equation (2.56) can be expressed in a compact form by introducing the following notation. Given a generic point P, represented in the reference frame $\langle b \rangle$ by the algebraic vector:

$$^b\boldsymbol{p} = \begin{bmatrix} x_b \\ y_b \\ z_b \end{bmatrix}$$

The same can be represented also with the form:

$$^b\overline{\boldsymbol{p}} = \begin{bmatrix} x_b \\ y_b \\ z_b \\ 1 \end{bmatrix} \tag{2.57}$$

Equation (2.57) is the representation of P in *homogeneous coordinates*. Equation (2.56) can thus be rewritten in the following matrix form:

$$^a\overline{\boldsymbol{p}} = {^a_b\boldsymbol{T}}{^b\overline{\boldsymbol{p}}} \tag{2.58}$$

where the 4×4 matrix:

$$^a_b\boldsymbol{T} = \begin{bmatrix} {^a_b\boldsymbol{R}} & {^aO_b} \\ 0 \;\; 0 \;\; 0 & 1 \end{bmatrix} \tag{2.59}$$

is the *transformation matrix*, containing all the information about the reference system (orientation and origin position).

Similarly, the following relationship holds:

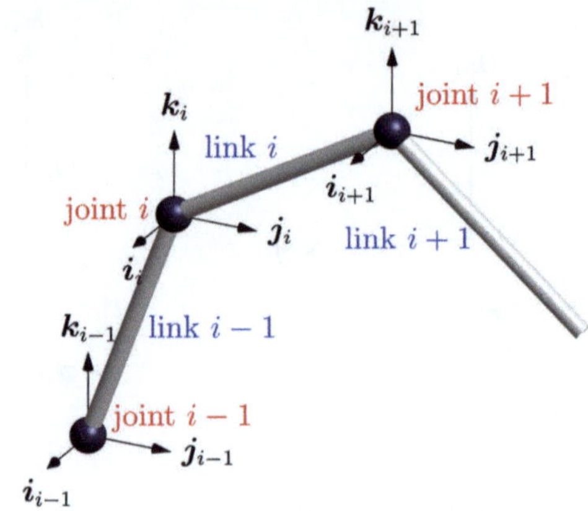

Fig. 2.7 Kinematic chain: joint, link and frame enumeration

$$^{b}\overline{p} = {}^{b}_{a}T\,{}^{a}\overline{p} \tag{2.60}$$

Comparing Eqs. (2.58) and (2.60) we obtain:

$$^{b}_{a}T = {}^{a}_{b}T^{-1} \tag{2.61}$$

In a multi-frame structure, the computation of the transformation matrix between two generic frames $\langle k \rangle$ and $\langle h \rangle$ can be done in the same fashion described in Sect. 2.1.2.6 for the rotation matrix.

2.1.3 Geometry of Robotics Structures

The robotics structure here considered is defined as a collection of rigid bodies (*links*) connected by mechanical devices (*joints*) having the purpose of limiting the degrees of freedom between adjacent links (Fig. 2.7). Such structures are known as *kinematic chain* and can be classified as:

- Open chain with serial (linear) topology
- Open chain with ramified topology
- Closed chain

In this work we will be considering only open chains.

The most common joints encountered in robotics are *rotational* and *translational*. The former allow only rotation around one fixed axis, while the latter permit the translation along one fixed direction.

Sometime it is possible to find joints with three degrees of freedom, for example when the joint must allow every orientation in the space. Multi degrees of freedom joints can be made up of a sequence of three rotational joints, each one allowing a rotation around a predefined axis (as for example the $\boldsymbol{k}, \boldsymbol{j}$ and $\boldsymbol{i}$ axis sequence). This kind of compound joint, despite its simplicity, may present singular configurations, such as the well known *gimbal lock*. The singularity issue does not exist in the *spherical joint*. Even though it may be of difficult practical realization, the spherical joint can be used to simulate floating structures.

The kinematics of joints will be considered more thoroughly in the next section.

2.1.3.1 Description of a Robotic Structure

In the past robotic literature, one of the main effort in describing a generic robotic structure was to produce certain regularity in the matrix forms. The Denavit-Hartenberg [1] is, for example, a very common parameterization, whose main advantage is that the configuration matrices have a specified form. This was relevant when involving hand-calculations. With the introduction of the modern numeric and symbolic computation, this is no more a concern, and an increased simplicity and flexibility in the descriptive language may supersede the constraints of the old computational difficulties.

Through the introduction of the *joint matrix* (described in Sect. 2.2.2), the position and orientation of the link frames are no more constrained by some particular requirements, like for example having the motion direction along the same axis. In our parameterization, the motion directions are completely specified within the joint matrix, and the frame placement across the structure can be done using the following methodology.

As noted above, we are considering open chains with ramified topology; we must divide the structure in several linear kinematic chains and enumerate each link with a notation that considers also the chain (branch) to which it belongs. The most important steps in positioning the link frames are hence the following (see also Fig. 2.8):

1. Define first the main base frame (*environment*). The main frame will be indicated with the index pair $\langle 1, 0 \rangle$.
2. Decompose the structure in a certain number of linear chains (*branch*); at least one chain will have a link connected to the environment frame through a joint: this branch will be indicated with the branch number 1.
3. Select a branch of the remaining structure connected to the branch 1: this will be the branch number 2.
4. Move forward until all the branches are selected.
5. For each branch, enumerate progressively its links, starting from the link that connects the branch to the remaining of the structure; with this notation, a generic link is described by an index pair $\langle i, j \rangle$, where i is the branch number and j is the link number within the same branch.
6. Enumerate the joints by assigning to each one the same index pair of the link they are preceding.
7. Apply to each link a frame, with orientation parallel to the main frame, and with the origin coinciding with the point where the preceding joint is attached to the

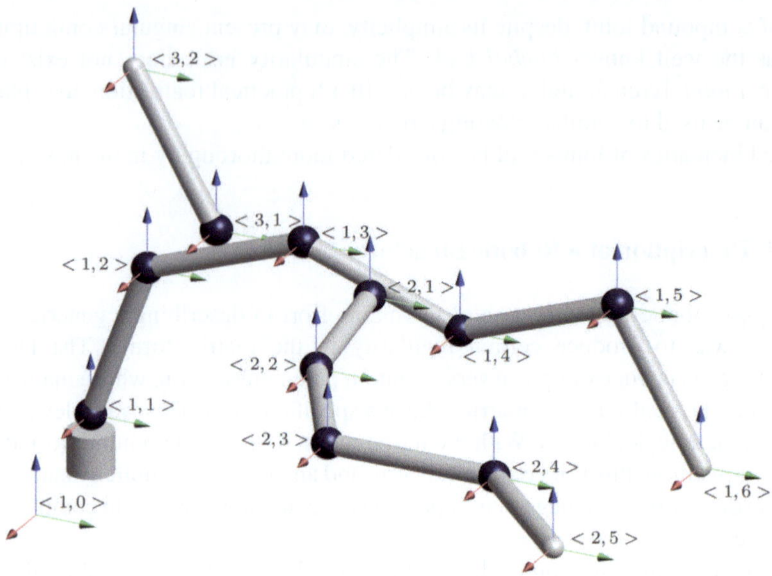

Fig. 2.8 Open kinematic chain with ramified topology

previous link; the frame will be identified by the same index pair $\langle i, j\rangle$ of the link, with i is the branch number and j the link number within the same branch.

8. Optionally, apply another frame to the last link of each branch, with orientation originally parallel to the main frame and the origin coinciding with the end-effector; the frame will be identified by the index pair $\langle i, n_i + 1\rangle$, where i is the branch number and n_i is the number of links of the same branch.

Figure 2.8 shows an application example of this methodology. It is important to observe the previous sequence: in particular, a branch cannot be connected to another having a greater index, otherwise the path toward the main frame would be lost. This is important when searching the shortest path between two links of different branches, during the computation of the transformation matrices.

In summary, to completely describe a generic open kinematic chain with branch topology we must specify:

1. number of branches
2. number of links per branch
3. point of connection of the joint 1 of each branch

and, for each link:

1. type of associated joint
2. point where the successive (outboard) joint is connected (end-effector in case of the last link)
3. dynamic parameters (mass, center of mass, inertia matrix).

2.2 Kinematics

This section introduces the fundamental concepts for the study of the motion (kinematics) of robots. In the following kinematic analysis the constrained motion introduced by the joints will be generalized in order to embrace a larger class of mechanical systems. A free body, for example, will be considered as a particular case of constraint, thus allowing an easy and intuitive approach to describe a generic mechanical system. After a summary of the necessary theoretical background, the section introduces the approach for describing a generic joint, and terminates with the solution of the direct kinematic problem (Jacobian).

2.2.1 Introduction to General Kinematics

Within this introduction to general kinematics we would like to make the reader familiar with the concepts used in our treatise. A complete review of general kinematics is outside the scope of this book and can be extensively found in the literature (as for example in McCarthy [2] or Goldstein [3]).

2.2.1.1 Vector Derivatives

The time derivative of the geometrical vector $\boldsymbol{\rho}$, computed with respect to the frame $\langle a\rangle$, is defined as the following geometrical vector:

$$\frac{d_a}{dt}\boldsymbol{\rho} \stackrel{\wedge}{=} \boldsymbol{i}_a\dot{x}_a + \boldsymbol{j}_a\dot{y}_a + \boldsymbol{k}_a\dot{z}_a \tag{2.62}$$

Similarly, for the same vector but in another frame $\langle b\rangle$ we have:

$$\frac{d_b}{dt}\boldsymbol{\rho} \stackrel{\wedge}{=} \boldsymbol{i}_b\dot{x}_b + \boldsymbol{j}_b\dot{y}_b + \boldsymbol{k}_b\dot{z}_b \tag{2.63}$$

Let's now find the relation between Eqs. (2.62) and (2.63):

$$\begin{aligned}\frac{d_a}{dt}\boldsymbol{\rho} &\stackrel{\wedge}{=} \frac{d_a}{dt}\left(\boldsymbol{i}_b x_b + \boldsymbol{j}_b y_b + \boldsymbol{k}_b z_b\right)\\ &= \boldsymbol{i}_b\dot{x}_b + \boldsymbol{j}_b\dot{y}_b + \boldsymbol{k}_b\dot{z}_b + \left(\frac{d_a}{dt}\boldsymbol{i}_b\right)x_b + \left(\frac{d_a}{dt}\boldsymbol{j}_b\right)y_b + \left(\frac{d_a}{dt}\boldsymbol{k}_b\right)z_b\end{aligned} \tag{2.64}$$

Note that, in general, projecting the (2.62) on the frame $\langle b\rangle$, we have:

$$^b\left(\frac{d_a}{dt}\boldsymbol{\rho}\right) \neq \frac{d_a}{dt}{}^b\boldsymbol{\rho} = \frac{d}{dt}{}^b\boldsymbol{\rho}. \tag{2.65}$$

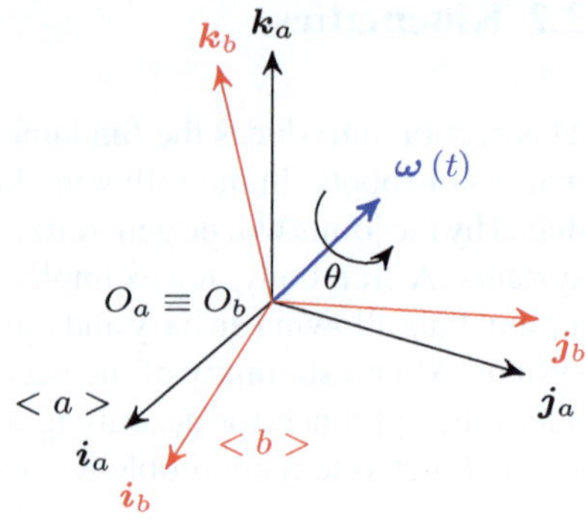

Fig. 2.9 Frames in relative motion

2.2.1.2 Angular Velocity

Let's consider two frames in relative motion as in Fig. 2.9. Since the matrix ${}^a_b\boldsymbol{R}(t)$ is time-dependent, it is possible to define as *angular velocity* of the frame $\langle b\rangle$ w.r.t. the frame $\langle a\rangle$ the vector $\boldsymbol{\omega}_{b/a}$, which, at any instant in time, provides the following information:

(a) the versor of $\boldsymbol{\omega}_{b/a}$ indicates the axis around which, at any instant, an observer sitting in $\langle a\rangle$ sees $\langle b\rangle$ rotating (w.r.t. $\langle a\rangle$);
(b) the component along its versor indicates the effective instantaneous angular velocity (measured in rad/s)

The angular velocity vector can be associated with the following differential form:

$$d\boldsymbol{\theta}_{b/a} = \boldsymbol{\omega}_{b/a} dt \tag{2.66}$$

Equation (2.66), however, does not coincide in general with any exact differential.[1]

It is possible to rewrite Eq. (2.64) by using the vector $\boldsymbol{\omega}_{b/a}$; at this aim we need the Poisson formulas (for its proof, see Goldstein [3]):

$$\begin{cases} \frac{d_a}{dt}\boldsymbol{i}_b = \boldsymbol{\omega}_{b/a} \times \boldsymbol{i}_b \\ \frac{d_a}{dt}\boldsymbol{j}_b = \boldsymbol{\omega}_{b/a} \times \boldsymbol{j}_b \\ \frac{d_a}{dt}\boldsymbol{k}_b = \boldsymbol{\omega}_{b/a} \times \boldsymbol{k}_b \end{cases} \tag{2.67}$$

Hence, Eq. (2.64) becomes:

$$\frac{d_a}{dt}\boldsymbol{\rho} = \frac{d_b}{dt}\boldsymbol{\rho} + \boldsymbol{\omega}_{b/a} \times \boldsymbol{\rho} \tag{2.68}$$

[1] A differential dQ is said to be exact, as contrasted with an inexact differential, if the differentiable function Q exists.

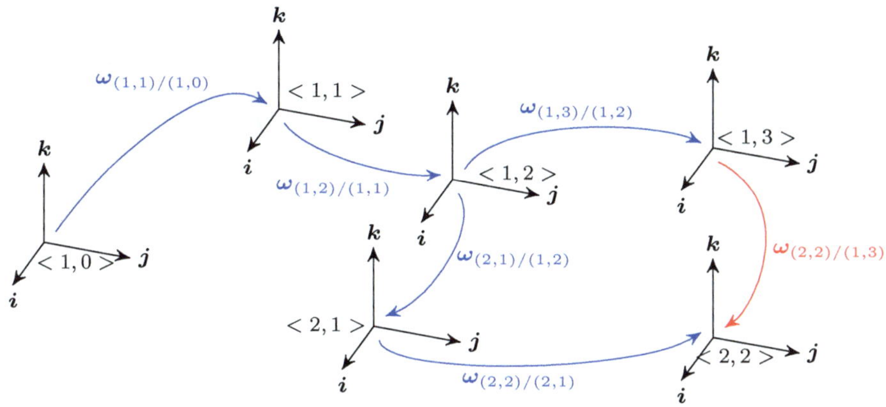

Fig. 2.10 Composition of angular velocities across different frames

Note that, if the module of the vector $\boldsymbol{\rho}$ is constant and $\boldsymbol{\rho}$ is integral with $\langle b \rangle$, Eq. (2.68) becomes:

$$\frac{d_a}{dt}\boldsymbol{\rho} = \boldsymbol{\omega}_{b/a} \times \boldsymbol{\rho} \tag{2.69}$$

The above relation is particularly useful in case of rigid bodies.

Some important properties of the angular velocity vector are the following:

1. $\boldsymbol{\omega}_{b/a} = -\boldsymbol{\omega}_{a/b}$
2. Given n frames, the angular velocity $\boldsymbol{\omega}_{k/h}$ of the a generic frame $\langle k \rangle$ with respect to any other frame $\langle h \rangle$ can be obtained by adding (vector sum) the successive angular velocities encountered in any path between k and h.

In the example of Fig. 2.10 we have:

$$\boldsymbol{\omega}_{(2,2)/(1,3)} = \boldsymbol{\omega}_{(2,2)/(2,1)} + \boldsymbol{\omega}_{(2,1)/(1,2)} - \boldsymbol{\omega}_{(1,3)/(1,2)}. \tag{2.70}$$

2.2.1.3 Time Derivatives for Points in Space

Let's consider now two reference systems $\langle a \rangle$ and $\langle b \rangle$, with the latter moving relatively to the former, with any general (either rotational and translational) motion. We define velocity of one point P (see Fig. 2.11), computed with respect to the frame $\langle a \rangle$, the geometric vector:

$$\boldsymbol{v}_{p/a} = \frac{d_a}{dt}(P - O_a) = \boldsymbol{i}_a \dot{x}_{p/a} + \boldsymbol{j}_a \dot{y}_{p/a} + \boldsymbol{k}_a \dot{z}_{p/a} \tag{2.71}$$

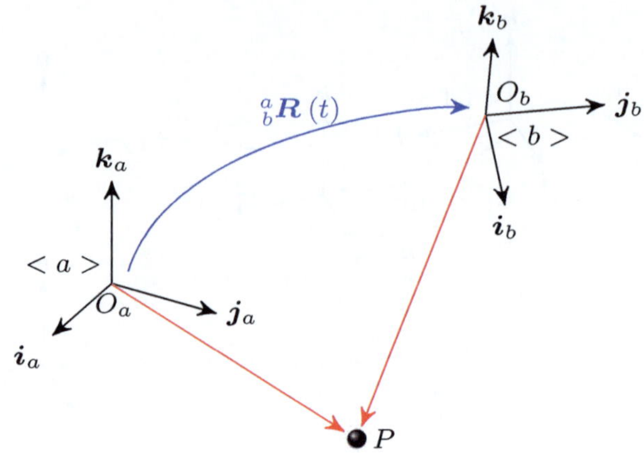

Fig. 2.11 Reference frames in relative motion with a generic point

Similarly:

$$\boldsymbol{v}_{p/b} = \frac{d_b}{dt}(P - O_b) = \boldsymbol{i}_b\dot{x}_{p/b} + \boldsymbol{j}_b\dot{y}_{p/b} + \boldsymbol{k}_b\dot{z}_{p/b} \tag{2.72}$$

Let's now find the relationship between Eqs. (2.71) and (2.72):

$$\boldsymbol{v}_{p/a} = \frac{d_a}{dt}(P - O_a) = \frac{d_a}{dt}(O_b - O_a) + \frac{d_a}{dt}(P - O_b) \tag{2.73}$$

Indicating with $\boldsymbol{v}_{b/a}$ the velocity of the origin of the frame $\langle b\rangle$ with respect to the frame $\langle a\rangle$, the previous relation becomes:

$$\boldsymbol{v}_{p/a} = \boldsymbol{v}_{b/a} + \frac{d_a}{dt}(P - O_b) \tag{2.74}$$

and, using Eq. (2.68) (with the opportune indices):

$$\boldsymbol{v}_{p/a} = \boldsymbol{v}_{b/a} + \frac{d_b}{dt}(P - O_b) + \boldsymbol{\omega}_{b/a} \times (P - O_b) \tag{2.75}$$

that is:

$$\boldsymbol{v}_{p/a} = \boldsymbol{v}_{b/a} + \boldsymbol{v}_{P/b} + \boldsymbol{\omega}_{b/a} \times (P - O_b) \tag{2.76}$$

In the common case where P is integral with $\langle b\rangle$, it happens to be $\boldsymbol{v}_{P/b} = 0$. Hence:

$$\boldsymbol{v}_{p/a} = \boldsymbol{v}_{b/a} + \boldsymbol{\omega}_{b/a} \times (P - O_b)\,. \tag{2.77}$$

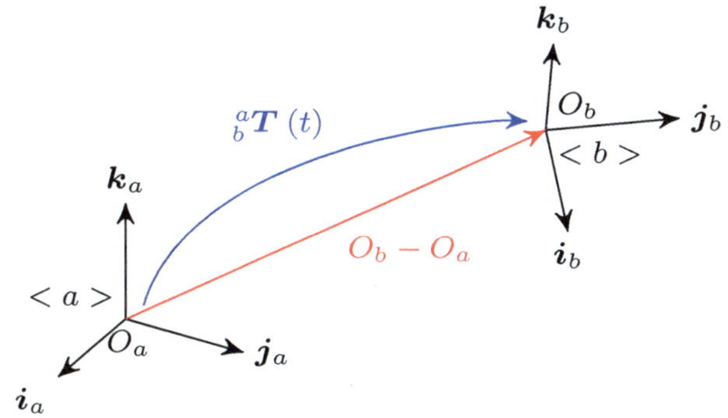

Fig. 2.12 Reference frames in relative motion

2.2.1.4 Generalized Velocity

In order to completely describe the relative motion between two reference frames (see Fig. 2.12), it is possible to organize in one vector the angular velocity and the velocity of the origin of the second frame with respect to the first. For example, in order to specify how the frame $\langle b \rangle$ moves with respect to $\langle a \rangle$, let's introduce the vector:

$$\dot{\boldsymbol{X}}_{b/a} \triangleq \begin{bmatrix} \boldsymbol{\omega}_{b/a} \\ \boldsymbol{v}_{b/a} \end{bmatrix} \tag{2.78}$$

where $\boldsymbol{v}_{b/a}$ is the velocity of the origin of the frame $\langle b \rangle$ with respect to $\langle a \rangle$ (velocity of the vector $O_b - O_a$). The vector $\dot{\boldsymbol{X}}_{b/a}$ is named generalized velocity of the frame $\langle b \rangle$ with respect to the frame $\langle a \rangle$. It can be projected on any frame. For example, projecting on the base frame $\langle 0 \rangle$, we have:

$$^0\dot{\boldsymbol{X}}_{b/a} \triangleq \begin{bmatrix} ^0\boldsymbol{\omega}_{b/a} \\ ^0\boldsymbol{v}_{b/a} \end{bmatrix} \tag{2.79}$$

Of course, the above relation holds even for a generic point integral with the frame $\langle b \rangle$. Let P be a point belonging to $\langle b \rangle$; we define *generalized velocity* of P with respect to $\langle a \rangle$ the quantity:

$$\dot{\boldsymbol{X}}_{P/a} \triangleq \begin{bmatrix} \boldsymbol{\omega}_{b/a} \\ \boldsymbol{v}_{P/a} \end{bmatrix} \tag{2.80}$$

with $\boldsymbol{v}_{P/a}$ defined by Eq. (2.77).

2.2.1.5 Derivative of the Rotation Matrix

With respect to Fig. 2.4, we now want to find a relationship between the derivative $^a_b\dot{\boldsymbol{R}}$ of the orientation matrix and the angular velocity vector $\boldsymbol{\omega}_{b/a}$. Recalling the (2.16):

$$ {}^a_b\boldsymbol{R} = \left[{}^a\boldsymbol{i}_b \; {}^a\boldsymbol{j}_b \; {}^a\boldsymbol{k}_b \right] \tag{2.81}$$

and deriving with respect to time we have:

$$\begin{aligned}
{}^a_b\dot{\boldsymbol{R}} &= \left[\tfrac{d}{dt}\,{}^a\boldsymbol{i}_b \; \tfrac{d}{dt}\,{}^a\boldsymbol{j}_b \; \tfrac{d}{dt}\,{}^a\boldsymbol{k}_b \right] \\
&= \left[{}^a\left(\tfrac{d_a}{dt}\boldsymbol{i}_b\right) \; {}^a\left(\tfrac{d_a}{dt}\boldsymbol{j}_b\right) \; {}^a\left(\tfrac{d_a}{dt}\boldsymbol{k}_b\right) \right] \\
&= \left[{}^a\left(\boldsymbol{\omega}_{b/a} \times \boldsymbol{i}_b\right) \; {}^a\left(\boldsymbol{\omega}_{b/a} \times \boldsymbol{j}_b\right) \; {}^a\left(\boldsymbol{\omega}_{b/a} \times \boldsymbol{k}_b\right) \right] \\
&= \left[\lfloor {}^a\boldsymbol{\omega}_{b/a}\times \rfloor \, {}^a\boldsymbol{i}_b \; \lfloor {}^a\boldsymbol{\omega}_{b/a}\times \rfloor \, {}^a\boldsymbol{j}_b \; \lfloor {}^a\boldsymbol{\omega}_{b/a}\times \rfloor \, {}^a\boldsymbol{k}_b \right] \\
&= \lfloor {}^a\boldsymbol{\omega}_{b/a}\times \rfloor \left[{}^a\boldsymbol{i}_b \; {}^a\boldsymbol{j}_b \; {}^a\boldsymbol{k}_b \right]
\end{aligned} \tag{2.82}$$

that is:

$$ {}^a_b\dot{\boldsymbol{R}} = \lfloor {}^a\boldsymbol{\omega}_{b/a}\times \rfloor \, {}^a_b\boldsymbol{R} \tag{2.83}$$

Substituting Eq. (2.29) within the previous Eq. (2.83) we obtain a dual formula:

$$ {}^a_b\dot{\boldsymbol{R}} = {}^a_b\boldsymbol{R} \left\lfloor {}^b\boldsymbol{\omega}_{b/a}\times \right\rfloor \tag{2.84}$$

Equations (2.83) and (2.84) are often employed for numerically evaluating the time evolution of ${}^a_b\boldsymbol{R}$; more precisely:

$$ {}^a_b\boldsymbol{R}\,(t+dt) = {}^a_b\boldsymbol{R}\,(t) + {}^a_b\dot{\boldsymbol{R}}dt. \tag{2.85}$$

2.2.2 Joint Kinematics

In general, the set of all the relative movements between two unconstrained rigid bodies forms a group, $\mathcal{G}$, consisting of all rotations and translations of $\Re^3$ [4]. A generic element of $\mathcal{G}$ may be represented by a matrix (see Fig. 2.12):

$$ {}^a_b\boldsymbol{T} = \begin{bmatrix} {}^a_b\boldsymbol{R} & \boldsymbol{L} \\ 0 & 1 \end{bmatrix} \in \mathcal{G}, \quad {}^a_b\boldsymbol{R} \in SO\,(3)\,, \quad \boldsymbol{L} \in \Re^3 \tag{2.86}$$

where $SO\,(3)$ is the group of the rotation matrices.

The group $\mathcal{G}$ is also said Special Euclidean group $SE\,(3)$. Its tangent space is isomorphic to $\mathcal{G} \times g$, where g is the space of the generalized velocities (*Lie* algebra). A generic element belonging to g can be formally represented using Eq. (2.84):

$$ {}^b\tilde{\boldsymbol{X}}_{b/a} = {}^a_b\boldsymbol{T}^{-1}\,{}^a_b\dot{\boldsymbol{T}} = \begin{bmatrix} {}^a_b\boldsymbol{R}^T & -\boldsymbol{L} \\ 0 & 1 \end{bmatrix} \begin{bmatrix} {}^a_b\dot{\boldsymbol{R}} & \dot{\boldsymbol{L}} \\ 0 & 0 \end{bmatrix} = \begin{bmatrix} {}^a_b\boldsymbol{R}^T\,{}^a_b\dot{\boldsymbol{R}} & {}^a_b\boldsymbol{R}^T\dot{\boldsymbol{L}} \\ 0 & 0 \end{bmatrix} = \begin{bmatrix} \lfloor {}^b\boldsymbol{\omega}_{b/a}\times \rfloor & {}^b\boldsymbol{v}_{b/a} \\ 0 & 0 \end{bmatrix} \tag{2.87}$$

The element ${}^{b}\tilde{\boldsymbol{X}}_{b/a}$ is the re-organization of the generalized velocity ${}^{b}\dot{\boldsymbol{X}}_{b/a}$ into a particular matrix form.

A generic joint, already introduced in Sect. 2.1.3, can be represented by a constraint relation in the tangent space $\mathcal{G} \times g$, involving the generalized velocity of the frame $\langle b \rangle$ with respect to $\langle a \rangle$:

$$\boldsymbol{A}\left(\boldsymbol{q}\right){}^{b}\dot{\boldsymbol{X}}_{b/a} = 0 \tag{2.88}$$

where $\boldsymbol{q}$ is the configuration. Joints of this type are known as *kinematic*, since they constrain the relative velocity.

Equation (2.88) means that ${}^{b}\dot{\boldsymbol{X}}_{b/a}$ belongs to the kernel of $\boldsymbol{A}\left(\boldsymbol{q}\right)$:

$${}^{b}\dot{\boldsymbol{X}}_{b/a} \in \Delta\left(q\right) = \ker\left[\boldsymbol{A}\left(\boldsymbol{q}\right)\right] \tag{2.89}$$

and, if the distribution $\Delta\left(\boldsymbol{q}\right)$ is integrable, the constraint introduced by the joint is known as *Holonomic*.

In case the reference frame is integral with at least one of the two bodies, the matrix $\Delta\left(\boldsymbol{q}\right) = \boldsymbol{A}$ is constant (independent of the configuration). This class of joints is called *Simple Kinematic Joints*. For this class, let's introduce a matrix $\boldsymbol{H}$, whose columns form a basis of the kernel of $\boldsymbol{A}$. All the solutions of Eq. (2.88) are:

$${}^{b}\dot{\boldsymbol{X}}_{b/a} = \boldsymbol{H}\boldsymbol{p}, \quad \boldsymbol{p} \in \Re^{r} \tag{2.90}$$

with $r = \dim\left(\ker\left[\boldsymbol{A}\right]\right)$, degrees of freedom of the joint. $\boldsymbol{H}$ is the Joint Matrix [5], while $\boldsymbol{p}$ is often named *quasi-velocity*. Table 2.1 shows some examples of joint matrices associated to commonly used joints.

2.2.2.1 Parameterization of Simple Kinematic Joints

The configuration of the joint is, in general, defined by the differential equation (2.87):

$${}^{a}_{b}\dot{\boldsymbol{T}} = {}^{a}_{b}\boldsymbol{T}\,{}^{b}\tilde{\boldsymbol{X}}_{b/a} \tag{2.91}$$

By using Eq. (2.90), the last can be re-written in the following form:

$${}^{a}_{b}\dot{\boldsymbol{T}} = {}^{a}_{b}\boldsymbol{T}\begin{bmatrix} \lfloor \boldsymbol{H}_{1}\boldsymbol{p}\times\rfloor & \boldsymbol{H}_{2}\boldsymbol{p} \\ 0 & 0 \end{bmatrix} \tag{2.92}$$

where $\boldsymbol{H}_1$ contains the first three rows of $\boldsymbol{H}$ and $\boldsymbol{H}_2$ the last three. Once the quasi-velocities $\boldsymbol{p}$ are known, Eq. (2.92) can be integrated to compute the time evolution of the transformation matrix $\boldsymbol{T}\left(t\right)$. However, Eq. (2.92) gives more information than necessary, since the movement is constrained to lie on a subgroup of $\mathcal{G}$. For example, let's consider an one degree of freedom joint ($r = 1$). In this case the joint matrix

Table 2.1 Matrices for commonly used joints

$\begin{bmatrix}0\\0\\1\\0\\0\\0\end{bmatrix}$	$\begin{bmatrix}0\\0\\0\\1\\0\\0\end{bmatrix}$	$\begin{bmatrix}0\\0\\1\\0\\0\\k\end{bmatrix}$	$\begin{bmatrix}1&0&0\\0&1&0\\0&0&1\\0&0&0\\0&0&0\\0&0&0\end{bmatrix}$
1 DOF Revolute joint (body z-axis)	1 DOF Prismatic joint (body x-axis)	1 DOF Screw joint (body z-axis)	3 DOF Spherical joint

$\boldsymbol{H} = \boldsymbol{h} \in \Re^6$ is made of only one column and the transformation matrix ${}^a_b\boldsymbol{T}$ can be parameterized using one parameter q_1:

$$ {}^a_b\boldsymbol{T}(q_1) = \begin{bmatrix} \boldsymbol{R}(q_1) & \boldsymbol{L}(q_1) \\ 0 & 1 \end{bmatrix} \tag{2.93} $$

where:

$$ \boldsymbol{R}(q_1) = e^{\lfloor \boldsymbol{h}_1 \times \rfloor q_1} \tag{2.94} $$

$$ \boldsymbol{L}(q_1) = \int_0^{q_1} e^{\lfloor \boldsymbol{h}_1 \times \rfloor \sigma} \boldsymbol{h}_2 d\sigma \tag{2.95} $$

with $\boldsymbol{h} = \begin{bmatrix} \boldsymbol{h}_1 \\ \boldsymbol{h}_2 \end{bmatrix}$.

Equation (2.94) can be easily derived by considering that vector $\boldsymbol{h}_1$ represents the direction of the rotation axis of the joint in the frame $\langle b \rangle$ and recalling the exponential representation of rotations (2.33).

Similarly, $\boldsymbol{h}_2$ is the direction of the translation axis of the joint in the reference frame $\langle b \rangle$. Therefore we can write:

$$ \frac{d\boldsymbol{L}}{dq_1} = \boldsymbol{R}(q_1)\boldsymbol{h}_2 \tag{2.96} $$

Equation (2.96), once integrated, results in Eq. (2.95).

In case $\boldsymbol{H}$ is made of $r > 1$ columns, if the joint is holonomic (and only in this case) its parameterization can be derived with the following procedure:

1. Associate with every column i a joint variable q_i
2. Compute the transformation matrix $\boldsymbol{T}_i$ for each column according to Eqs. (2.93), (2.94) and (2.95)
3. Compute the final transformation matrix by multiplying the r matrices as follows:

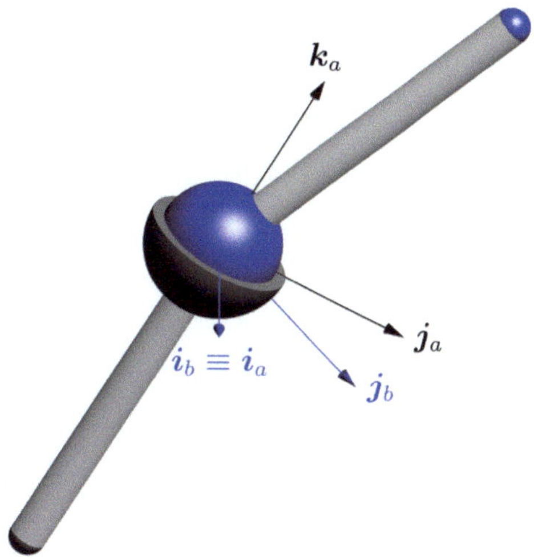

Fig. 2.13 Kinematics of a spherical joint

$$\boldsymbol{T}(\boldsymbol{q}) = \boldsymbol{T}_r(q_r) \ldots \boldsymbol{T}_2(q_2)\,\boldsymbol{T}_1(q_1)\,. \tag{2.97}$$

2.2.2.2 Example: Spherical Joint

The spherical joint (Fig. 2.13), also known as *ball-socket* joint, is characterized by a three-column joint matrix:

$$\boldsymbol{H}_{sp} = \begin{bmatrix} 1 & 0 & 0 \\ 0 & 1 & 0 \\ 0 & 0 & 1 \\ 0 & 0 & 0 \\ 0 & 0 & 0 \\ 0 & 0 & 0 \end{bmatrix} \tag{2.98}$$

Let $\boldsymbol{p} = \begin{bmatrix} p_x & p_y & p_z \end{bmatrix}^T$. From Eq. (2.90) we have:

$${}^{b}\dot{\boldsymbol{X}}_{b/a} = \boldsymbol{H}\boldsymbol{p} = \begin{bmatrix} p_x \\ p_y \\ p_z \\ 0 \\ 0 \\ 0 \end{bmatrix} \tag{2.99}$$

This means that the frame $\langle b \rangle$ can move relatively to $\langle a \rangle$ with any rotation around its coordinate axis.

The parameterization can be obtained by first applying Eqs. (2.94) and (2.95) to every column:

$$\boldsymbol{R}_1(q_1) = e^{\left[\begin{bmatrix}1\\0\\0\end{bmatrix}\times\right]q_1} = \begin{bmatrix}1 & 0 & 0\\ 0 & \cos(q_1) & -\sin(q_1)\\ 0 & \sin(q_1) & \cos(q_1)\end{bmatrix} \tag{2.100}$$

$$\boldsymbol{L}_1(q_1) = \begin{bmatrix}0\\0\\0\end{bmatrix} \tag{2.101}$$

$$\boldsymbol{R}_2(q_2) = e^{\left[\begin{bmatrix}0\\1\\0\end{bmatrix}\times\right]q_2} = \begin{bmatrix}\cos(q_2) & 0 & \sin(q_2)\\ 0 & 1 & 0\\ -\sin(q_2) & 0 & \cos(q_2)\end{bmatrix} \tag{2.102}$$

$$\boldsymbol{L}_2(q_2) = \begin{bmatrix}0\\0\\0\end{bmatrix} \tag{2.103}$$

$$\boldsymbol{R}_3(q_3) = e^{\left[\begin{bmatrix}0\\0\\1\end{bmatrix}\times\right]q_3} = \begin{bmatrix}\cos(q_3) & -\sin(q_3) & 0\\ \sin(q_3) & \cos(q_3) & 0\\ 0 & 0 & 1\end{bmatrix} \tag{2.104}$$

$$\boldsymbol{L}_3(q_3) = \begin{bmatrix}0\\0\\0\end{bmatrix} \tag{2.105}$$

Finally, the transformation (configuration) matrix of the joint can be computed using Eq. (2.97) (see also Fig. 2.13):

$$\begin{aligned}{}^{a}_{b}\boldsymbol{T}(\boldsymbol{q}) &= \boldsymbol{T}_3(q_3)\,\boldsymbol{T}_2(q_2)\,\boldsymbol{T}_1(q_1)\\ &= \begin{bmatrix}\cos(q3)\cos(q2) & -\sin(q3)\cos(q1)+\cos(q3)\sin(q2)\sin(q1) & \sin(q3)\sin(q1)+\cos(q3)\sin(q2)\cos(q1) & 0\\ \sin(q3)\cos(q2) & \cos(q3)\cos(q1)+\sin(q3)\sin(q2)\sin(q1) & -\cos(q3)\sin(q1)+\sin(q3)\sin(q2)\cos(q1) & 0\\ -\sin(q2) & \cos(q2)\sin(q1) & \cos(q2)\cos(q1) & 0\\ 0 & 0 & 0 & 1\end{bmatrix}\end{aligned} \tag{2.106}$$

Note that the rotational part of Eq. (2.106) coincides with Eq. (2.42). As a matter of fact, with this choice, the parameters representing the joint are the Roll, Pitch and Yaw angles.

2.2.2.3 Example: Screw Joint

Within the joint matrix, it is possible to specify both a rotational and translational axis for the same joint variable (the same degree of freedom). Let's consider, for example, a joint matrix $\boldsymbol{H}$ like the following:

$$\boldsymbol{H}_{sc} = \begin{bmatrix} 1 \\ 0 \\ 0 \\ 1 \\ 0 \\ 0 \end{bmatrix} \tag{2.107}$$

A variation of q_1 results in a rotation around the iaxis and, in the same time, a translation along the same axis, similarly to the movement of a screw. The associated matrices are:

$$\boldsymbol{R}_1(q_1) = e^{\left[\left[\begin{smallmatrix} 1 \\ 0 \\ 0 \end{smallmatrix}\right] \times q_1\right]} = \begin{bmatrix} 1 & 0 & 0 \\ 0 & \cos(q_1) & -\sin(q_1) \\ 0 & \sin(q_1) & \cos(q_1) \end{bmatrix} \tag{2.108}$$

$$\boldsymbol{L}_{sc}(q_1) = \int_0^{q_1} \boldsymbol{R}_{sc}(\sigma) \begin{bmatrix} 1 \\ 0 \\ 0 \end{bmatrix} d\sigma = \int_0^{q_1} \begin{bmatrix} 1 \\ 0 \\ 0 \end{bmatrix} d\sigma = \begin{bmatrix} q_1 \\ 0 \\ 0 \end{bmatrix} \tag{2.109}$$

Hence:

$$\boldsymbol{T}_{sc}(q_1) = \begin{bmatrix} 1 & 0 & 0 & q_1 \\ 0 & \cos(q_1) & -\sin(q_1) & 0 \\ 0 & \sin(q_1) & \cos(q_1) & 0 \\ 0 & 0 & 0 & 1 \end{bmatrix}. \tag{2.110}$$

2.2.2.4 Kinematic Equation of Simple Joints

The goal is now to find a relation between the quasi-velocities and the time derivative of the variables used to parameterize the joint. At this aim, let's consider a simple joint described by the matrix:

$$H = \begin{bmatrix} \boldsymbol{h}_{11} & \cdots & \boldsymbol{h}_{1r} \\ \boldsymbol{h}_{21} & \cdots & \boldsymbol{h}_{2r} \end{bmatrix} \in \Re^{6\times r} \tag{2.111}$$

parameterized as in Sect. 2.2.2.1. We define *kinematic equation* the following relation:

$$\dot{\boldsymbol{q}} = \boldsymbol{\Gamma}(\boldsymbol{q})\boldsymbol{p} \tag{2.112}$$

where the matrix $\boldsymbol{\Gamma}(\boldsymbol{q})$ is defined with the following recursive algorithm:

1. For $j = 1 \dots r$ let the matrices R_j and L_j be defined as:

$$\mathsf{R}_j(q_j \dots q_1) = R_j(q_j)\,\mathsf{R}_{j-1}(q_{j-1} \dots q_1), \quad \mathsf{R}(0) = I \tag{2.113}$$

$$\mathsf{L}_j(q_j \dots q_1) = R_j(q_j)\,\mathsf{L}_{j-1}(q_{j-1} \dots q_1) + \mathsf{L}_j(q_j), \quad \mathsf{L}(0) = I \tag{2.114}$$

   with $R_j(q_j)$ given by Eq. (2.94).
2. Let $\boldsymbol{B}$ be defined as:

$$\boldsymbol{B}(\boldsymbol{q}) = \begin{bmatrix} \boldsymbol{b}_{11} \cdots \boldsymbol{b}_{1r} \\ \boldsymbol{b}_{21} \dots \boldsymbol{b}_{2r} \end{bmatrix} \tag{2.115}$$

   where:

$$\lfloor \boldsymbol{b}_{1i} \times \rfloor = \mathsf{R}_{i-1}^T \lfloor \boldsymbol{h}_{1i} \times \rfloor \mathsf{R}_{i-1} \tag{2.116}$$

$$\boldsymbol{b}_{2i} = \mathsf{R}_{i-1}^T \lfloor \boldsymbol{h}_{1i} \times \rfloor \mathsf{L}_{i-1} + \mathsf{R}_{i-1}\boldsymbol{h}_{2i} \tag{2.117}$$

3. Compute $\boldsymbol{\Gamma}(\boldsymbol{q})$ as:

$$\boldsymbol{\Gamma}(\boldsymbol{q}) = \boldsymbol{B}^*(\boldsymbol{q})\,\boldsymbol{H} \tag{2.118}$$

   where $\boldsymbol{B}^*(\boldsymbol{q})$ denotes a right-inverse of $\boldsymbol{B}(\boldsymbol{q})$.

Proof.
Let's compute the derivative of the transformation matrix as a function of the derivatives $\dot{\boldsymbol{q}}$:

$$\begin{aligned} {}^a_b\dot{\boldsymbol{T}} &= \sum_{i=1}^{r} \frac{\partial\, {}^a_b\boldsymbol{T}(\boldsymbol{q})}{\partial q_1}\dot{q}_1 \\ &= \sum_{i=1}^{r} \left[\boldsymbol{T}_r(q_r) \dots \boldsymbol{T}_{i+1}(q_{i+1}) \frac{\partial \boldsymbol{T}_i(q_i)}{\partial q_i} \boldsymbol{T}_{i-1}(q_{i-1}) \dots \boldsymbol{T}_1(q_1) \right] \dot{q}_i \end{aligned} \tag{2.119}$$

Pre-multiplying by ${}^a_b\boldsymbol{T}(\boldsymbol{q})^{-1} = \{\boldsymbol{T}_r(q_r) \dots \boldsymbol{T}_2(q_2)\,\boldsymbol{T}_1(q_1)\}^{-1}$ and considering the Eq. (2.87) we obtain:

$$
\begin{aligned}
{}_b^a\boldsymbol{T}(\boldsymbol{q})^{-1}\,{}_b^a\dot{\boldsymbol{T}} &= {}^b\tilde{\boldsymbol{X}}_{b/a}\\
&= \sum_{i=1}^{r} \boldsymbol{T}_1(q_1)^{-1}\,\boldsymbol{T}_2(q_2)^{-1}\,\ldots\,\boldsymbol{T}_r(q_r)^{-1}\\
&\quad \left[\boldsymbol{T}_r(q_r)\,\ldots\,\boldsymbol{T}_{i+1}(q_{i+1})\,\frac{\partial \boldsymbol{T}_i(q_i)}{\partial q_i}\,\boldsymbol{T}_{i-1}(q_{i-1})\,\ldots\,\boldsymbol{T}_1(q_1)\right]\dot{q}_i\\
&= \sum_{i=1}^{r}\left[\boldsymbol{T}_1(q_1)^{-1}\,\boldsymbol{T}_2(q_2)^{-1}\,\ldots\,\boldsymbol{T}_i(q_i)^{-1}\,\frac{\partial \boldsymbol{T}_i(q_i)}{\partial q_i}\,\boldsymbol{T}_{i-1}(q_{i-1})\,\ldots\,\boldsymbol{T}_1(q_1)\right]\dot{q}_i\\
&= \sum_{i=1}^{r}\left[\boldsymbol{T}_{i-1}(q_{i-1})\,\ldots\,\boldsymbol{T}_1(q_1)\right]^{-1}\boldsymbol{T}_i(q_i)^{-1}\,\frac{\partial \boldsymbol{T}_i(q_i)}{\partial q_i}\,\boldsymbol{T}_{i-1}(q_{i-1})\,\ldots\,\boldsymbol{T}_1(q_1)\,\dot{q}_i
\end{aligned}
\tag{2.120}
$$

From Eq. (2.92) it is possible to show that:

$$
\boldsymbol{T}_i(q_i)^{-1}\,\frac{\partial \boldsymbol{T}_i(q_i)}{\partial q_i} = \begin{bmatrix} \lfloor \boldsymbol{h}_{1i}\times\rfloor & \boldsymbol{h}_{2i} \\ 0 & 0 \end{bmatrix} \tag{2.121}
$$

Then, defining the matrix

$$
\boldsymbol{U}_j(q_j\,\ldots\,q_1) = \boldsymbol{T}_j(q_j)\,\ldots\,\boldsymbol{T}_1(q_1) \tag{2.122}
$$

the (2.120) becomes:

$$
{}^b\tilde{\boldsymbol{X}}_{b/a} = \sum_{i=1}^{r} \boldsymbol{U}_{i-1}^{-1} \begin{bmatrix} \lfloor \boldsymbol{h}_{1i}\times\rfloor & \boldsymbol{h}_{2i} \\ 0 & 0 \end{bmatrix} \boldsymbol{U}_{i-1}\,\dot{q}_i \tag{2.123}
$$

It is easy to show that the matrix $\boldsymbol{U}_j$ has the form:

$$
\boldsymbol{U}_j = \begin{bmatrix} \mathsf{R}_j(q_j\,\ldots\,q_1) & \mathsf{L}_j(q_j\,\ldots\,q_1) \\ 0 & 1 \end{bmatrix} \tag{2.124}
$$

with R and L defined as in Eqs. (2.113) and (2.114). Hence, Eq. (2.123) becomes:

$$
\tilde{X} = \sum_{i=1}^{r} \begin{bmatrix} \mathsf{R}_{i-1}^T\,\lfloor \boldsymbol{h}_{1i}\times\rfloor\,\mathsf{R}_{i-1} & \mathsf{R}_{i-1}^T\,\lfloor \boldsymbol{h}_{1i}\times\rfloor\,\mathsf{L}_{i-1} + \mathsf{R}_{i-1}\boldsymbol{h}_{2i} \\ 0 & 0 \end{bmatrix} \dot{q}_i \tag{2.125}
$$

As already seen in Eq. (2.29), the matrix $\mathsf{R}_{i-1}^T\,\lfloor \boldsymbol{h}_{ij}\times\rfloor\,\mathsf{R}_{i-1}$ is still anti-symmetric. Is then possible to define a vector $\boldsymbol{b}_{1i} \in \Re^3$ such as:

$$
\lfloor \boldsymbol{b}_{1i}\times\rfloor = \mathsf{R}_{i-1}^T\,\lfloor \boldsymbol{h}_{1i}\times\rfloor\,\mathsf{R}_{i-1} \tag{2.126}
$$

Let's define also the vector:

$$\boldsymbol{b}_{1j} = \mathsf{R}_{i-1}^T \lfloor \boldsymbol{h}_{1i} \times \rfloor \mathsf{L}_{i-1} + \mathsf{R}_{i-1}^T \boldsymbol{h}_{2i} \tag{2.127}$$

so that Eq. (2.125) finally becomes:

$${}^b\tilde{\boldsymbol{X}}_{b/a} = \sum_{i=1}^{r} \begin{bmatrix} \lfloor \boldsymbol{b}_{1i} \times \rfloor & \boldsymbol{b}_{2i} \\ 0 & 0 \end{bmatrix} \dot{q}_i \tag{2.128}$$

It is possible to reorganize Eq. (2.128) in a compact form, substituting the anti-symmetric matrices with the relative vectors:

$${}^b\dot{\boldsymbol{X}}_{b/a} = \sum_{i=1}^{r} \begin{bmatrix} \boldsymbol{b}_{1i} \\ \boldsymbol{b}_{2i} \end{bmatrix} \dot{q}_i = \boldsymbol{B}(\boldsymbol{q})\,\dot{\boldsymbol{q}} \tag{2.129}$$

with $\boldsymbol{B}(\boldsymbol{q})$ defined as in Eqs. (2.115), (2.116) and (2.117). Recalling Eq. (2.90) we then have:

$$\boldsymbol{B}(\boldsymbol{q})\,\dot{\boldsymbol{q}} = \boldsymbol{H}\boldsymbol{p} \tag{2.130}$$

and:

$$\boldsymbol{\Gamma}(\boldsymbol{q}) = \boldsymbol{B}^*(\boldsymbol{q})\,\boldsymbol{H}. \tag{2.131}$$

2.2.2.5 Compound Joints

Not all the joints can be considered simple. In several cases, its action must be regarded as a sequence of simple joints, as for example the joint with roll-pitch-yaw angles of Fig. 2.14. Joints of this model will be named *compound*.

In a more general view, a compound joint can be characterized by the relative motion of a sequence of k frames such as the relative motion between frames is constrained by a simple joint. Each of the k simple joints is characterized by:

- a joint matrix $\boldsymbol{H}_i$ having r_i columns[2];
- a vector of parameters $\boldsymbol{q}_i$ of dimension r_i;
- a quasivelocity vector $\boldsymbol{p}_i$ of dimension r_i;
- a kinematic matrix $\boldsymbol{\Gamma}_i$ of dimension $r_i \times r_i$

With the aforementioned description, the compound joint can be considered as one unique device having:

[2] we will consider only the case $r_i = 1$, that is only compound joints made of successions of one DOF simple joints can be described.

- a configuration vector:

$$\boldsymbol{q} = \begin{bmatrix} \boldsymbol{q}_1 \\ \vdots \\ \boldsymbol{q}_k \end{bmatrix}$$

- a quasivelocity vector:

$$\boldsymbol{p} = \begin{bmatrix} \boldsymbol{p}_1 \\ \vdots \\ \boldsymbol{p}_k \end{bmatrix}$$

- a kinematic matrix such as:

$$\dot{\boldsymbol{q}} = \begin{bmatrix} \boldsymbol{\Gamma}_1(\boldsymbol{q}_1) & \mathbf{0} & \cdots & \mathbf{0} \\ \mathbf{0} & \boldsymbol{\Gamma}_2(\boldsymbol{q}_2) & \cdots & \mathbf{0} \\ \vdots & \vdots & \ddots & \vdots \\ \mathbf{0} & \mathbf{0} & \cdots & \boldsymbol{\Gamma}_k(\boldsymbol{q}_k) \end{bmatrix} \boldsymbol{p}$$

- a configuration matrix:

$$\boldsymbol{T}(\boldsymbol{q}) = \boldsymbol{T}_k(\boldsymbol{q}_i)\, \boldsymbol{T}_{k-1}(\boldsymbol{q}_{k-1}) \ldots \boldsymbol{T}_1(\boldsymbol{q}_1)$$

where we assume that the (simple) joins number 1 is the outer one.

In order to compute the global joint matrix, let's consider the simplified case with $r_i = 1,\ \forall i$. We can then determine the global joint matrix with the following recursive procedure, similar to the one used for computing the kinematic matrix of the simple joints.

Let's consider a compound joint made of k simple joints, each one of one degree of freedom and having a joint matrix $\boldsymbol{h}_i = \left[\, \boldsymbol{h}_{1i}\ \boldsymbol{h}_{21} \,\right]^T$. Then, the global joint matrix H is given by:

$$\boldsymbol{H}(\boldsymbol{q}) = \begin{bmatrix} \boldsymbol{h}_{11} & \cdots & \boldsymbol{h}_{1r} \\ \boldsymbol{h}_{21} & \cdots & \boldsymbol{h}_{2r} \end{bmatrix} \tag{2.132}$$

where:

$$\lfloor \boldsymbol{h}_{1i} \times \rfloor = \mathsf{R}_{i-1}^T \lfloor \boldsymbol{h}_{1i} \times \rfloor \mathsf{R}_{i-1} \tag{2.133}$$

$$\boldsymbol{h}_{21} = \mathsf{R}_{i-1}^T \lfloor \boldsymbol{h}_{1i} \times \rfloor \mathsf{L}_{i-1} + \mathsf{R}_{i-1}^T \boldsymbol{h}_{2i} \tag{2.134}$$

and:

$$\mathsf{R}_j(q_j \ldots q_1) = R_j(q_j)\, \mathsf{R}_{j-1}(q_{j-1} \ldots q_1), \quad \mathsf{R}(0) = I \tag{2.135}$$

$$\mathsf{L}_j\left(q_j\ldots q_1\right) = R_j\left(q_j\right)\mathsf{L}_{j-1}\left(q_{j-1}\ldots q_1\right) + \mathsf{L}_j\left(q_j\right), \quad L\left(0\right) = I \tag{2.136}$$

The proof will not be reported.

In conclusion, similarly to the simple joints, a compound joint can be described by a joint matrix $\boldsymbol{H}$ such as Eq. (2.90) still holds:

$$^{b}\dot{\boldsymbol{X}}_{b/a} = \boldsymbol{H}\boldsymbol{p}, \quad \boldsymbol{p} \in \Re^{r} \tag{2.137}$$

In this case (and in the hypothesis that $r_i = 1, \ \forall i$) the quasi-velocities coincide with the derivatives of the joint variables, and the matrix $\boldsymbol{H}$ performs a projection, in the outer frame, of the velocities at the single component joints.

2.2.2.6 Example: Compound Joint with Roll-Pitch-Yaw Angles

Let's consider a joint made of a succession of three one-DOF rotational simple joints, having the rotational directions along, respectively, the axis $\boldsymbol{i}, \boldsymbol{j}$ and $\boldsymbol{k}$ (starting from the outer frame, see Fig. 2.14). Organizing the joint matrices into a unique matrix we have:

$$\boldsymbol{H} = \left[\boldsymbol{h}_1\ \boldsymbol{h}_2\ \boldsymbol{h}_3\right] = \begin{bmatrix} 1\,0\,0 \\ 0\,1\,0 \\ 0\,0\,1 \\ 0\,0\,0 \\ 0\,0\,0 \\ 0\,0\,0 \end{bmatrix} \tag{2.138}$$

The computation of $\boldsymbol{H}$ using Eqs. (2.132) through (2.136) gives the following result:

$$\boldsymbol{H} = \begin{bmatrix} 1 & 0 & -sin\left(q_2\right) \\ 0 & \cos\left(q_1\right) & sin\left(q_1\right)\cos\left(q_2\right) \\ 0 & -sin\left(q_1\right) & \cos\left(q_1\right)\cos\left(q_2\right) \\ 0 & 0 & 0 \\ 0 & 0 & 0 \\ 0 & 0 & 0 \end{bmatrix} \tag{2.139}$$

while the configuration matrix is still given by Eq. (2.106).

The application of Eq. (2.138) gives:

$$^{b}\dot{\boldsymbol{X}}_{b/a} = \boldsymbol{H}\begin{bmatrix} p_1 \\ p_2 \\ p_3 \end{bmatrix} = \begin{bmatrix} p_1 \\ p_2\cos\left(q_1\right) + p_3 sin\left(q_1\right)\cos\left(q_2\right) \\ -p_2 sin\left(q_1\right) + p_3\cos\left(q_1\right)\cos\left(q_2\right) \\ 0 \\ 0 \\ 0 \end{bmatrix} \tag{2.140}$$

Fig. 2.14 Example of compound roll-pith-yaw joint

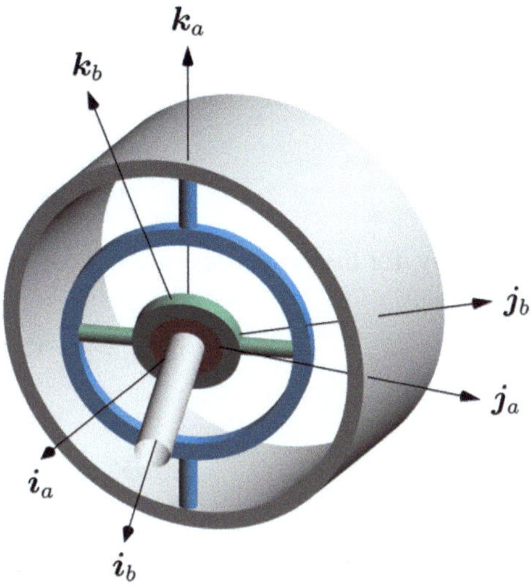

It is easy to verify that the first half of the matrix is the sum of the projections of the three angular velocities on the outer frame $\langle b \rangle$.

2.2.3 Kinematics of Robotic Systems

In the previous sections we have introduced the necessary background to solve the *direct kinematic problem*, i.e. to compute the generalized velocity of every point of the structure for a particular value of the quasi-velocity vector $\boldsymbol{p}$.

To simplify the study let's consider first a linear chain (one branch only, indicating with $\langle 0 \rangle$ the base frame), made of k links e k joints, each one characterized by:

- r_i degrees of freedom
- joint matrix $\boldsymbol{H}_i \in \Re^{6\times r_i}$
- configuration vector $\boldsymbol{q}_i \in \Re^{r_i}$
- quasivelocity vector $\boldsymbol{p}_i \in \Re^{r_i}$

Let c_i be a point of a generic link i and ${}^i\dot{\boldsymbol{X}}_{i/0}$ its generalized velocity w.r.t. the base frame as defined in Eq. (2.78), projected on the frame of the same link; relatively to the generalized velocity of the origin O_i of the frame $\langle i \rangle$ we have:

$$
\begin{aligned}
{}^i\dot{\boldsymbol{X}}_{c_i/0} &= \begin{bmatrix} {}^i\boldsymbol{\omega}_{c_i/0} \\ {}^i\boldsymbol{v}_{c_i/0} \end{bmatrix} = \begin{bmatrix} {}^i\boldsymbol{\omega}_{i/0} \\ {}^i\boldsymbol{v}_{i/0} + {}^i\boldsymbol{\omega}_{i/0} \times {}^i\boldsymbol{r}_{O_i c_i} \end{bmatrix} \\
&= \begin{bmatrix} \boldsymbol{I} & \boldsymbol{0} \\ -\lfloor {}^i\boldsymbol{r}_{O_i c_i} \times \rfloor & \boldsymbol{I} \end{bmatrix} \begin{bmatrix} {}^i\boldsymbol{\omega}_{i/0} \\ {}^i\boldsymbol{v}_{i/0} \end{bmatrix} = \phi\left({}^i\boldsymbol{r}_{O_i c_i}\right) {}^i\dot{\boldsymbol{X}}_{i/0}
\end{aligned}
\tag{2.141}
$$

where:

$$\phi(\boldsymbol{r}) = \begin{bmatrix} \boldsymbol{I} & \boldsymbol{0} \\ -\lfloor \boldsymbol{r}\times \rfloor & \boldsymbol{I} \end{bmatrix} \tag{2.142}$$

$$^{i}\boldsymbol{r}_{O_i c_i} = \overline{O_i c_i} \tag{2.143}$$

Similarly, for the properties of the cross product, we have:

$$^{i}\dot{\boldsymbol{X}}^{T}_{c_i/0} = {}^{i}\dot{\boldsymbol{X}}^{T}_{i/0}\phi^{*}\left({}^{i}\boldsymbol{r}_{O_i c_i}\right) \tag{2.144}$$

with:

$$\phi^{*}(\boldsymbol{r}) = \begin{bmatrix} \boldsymbol{I} & \lfloor \boldsymbol{r}\times \rfloor \\ \boldsymbol{0} & \boldsymbol{I} \end{bmatrix} \tag{2.145}$$

It is then possible to express, recursively, the generalized velocity of the frame integral with the generic link i with the relation:

$$^{i}\dot{\boldsymbol{X}}_{i/0} = \begin{bmatrix} {}^{i}_{i-1}\boldsymbol{R} & \boldsymbol{0} \\ \boldsymbol{0} & {}^{i}_{i-1}\boldsymbol{R} \end{bmatrix} \phi\left({}^{i-1}\boldsymbol{r}^{oj}_{i-1}\right) {}^{i-1}\dot{\boldsymbol{X}}_{i-1/0} + \boldsymbol{H}_{i}\boldsymbol{p}_{i} \tag{2.146}$$

where $\boldsymbol{r}^{oj}_{i-1}$ is the vector joining the origin of frame $\langle i-1\rangle$ with the connection point of the joint i (Fig. 2.15) and $\boldsymbol{p}_i$ is the associated quasi-velocity vector. Note that the quantity $\boldsymbol{H}_i\boldsymbol{p}_i$ gives, according to Eq. (2.90), the generalized velocity introduced by the joint and expressed in the "exit" frame.[3] Introducing the matrix:

$${}^{i}_{i-1}\boldsymbol{\theta}(\boldsymbol{r}) = \begin{bmatrix} {}^{i}_{i-1}\boldsymbol{R} & \boldsymbol{0} \\ \boldsymbol{0} & {}^{i}_{i-1}\boldsymbol{R} \end{bmatrix} \phi(\boldsymbol{r}) \tag{2.147}$$

Eq. (2.146) becomes:

$$^{i}\dot{\boldsymbol{X}}_{i/0} = {}^{i}_{i-1}\boldsymbol{\theta}\left({}^{i-1}\boldsymbol{r}^{oj}_{i-1}\right) {}^{i-1}\dot{\boldsymbol{X}}_{i-1/0} + \boldsymbol{H}_{i}\boldsymbol{p}_{i} \tag{2.148}$$

Note that ${}^{i}_{i-1}\boldsymbol{\theta}(\boldsymbol{r})$ is the matrix that transforms the generalized velocity of the frame $\langle i-1\rangle$ (expressed in the same frame $\langle i-1\rangle$) into the generalized velocity of the point integral with the frame $\langle i-1\rangle$ and specified by $\boldsymbol{r}$ (the last velocity is expressed in the frame $\langle i\rangle$).

[3] For example, with reference to Fig. 2.13, this quantity represents the generalized velocity of the frame $\langle b\rangle$ w.r.t. $\langle a\rangle$, expressed in the frame $\langle b\rangle$.

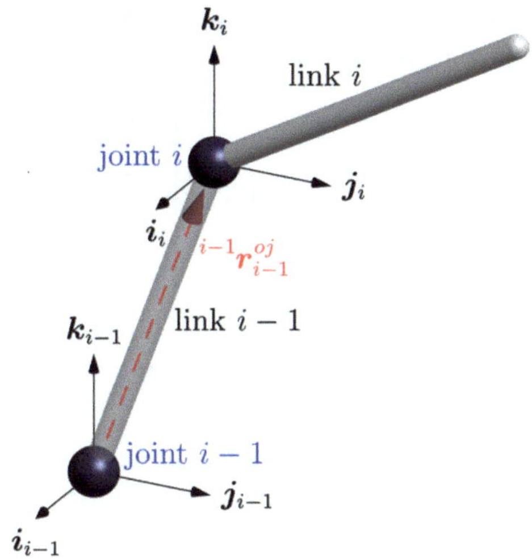

Fig. 2.15 Outboard joint

Equation (2.148) is already sufficient to recursively express all the different generalized velocities $\dot{\boldsymbol{X}}_{i/0}$ of every frame of the robot. However, it is convenient to expand the recursive process into a unique global form. At this aim, let's introduce the following vectors:

$$\boldsymbol{V}_0 = \begin{bmatrix} {}^1\dot{\boldsymbol{X}}_{1/0} \\ {}^2\dot{\boldsymbol{X}}_{2/0} \\ \vdots \\ {}^k\dot{\boldsymbol{X}}_{k/0} \end{bmatrix}, \quad \boldsymbol{p} = \begin{bmatrix} \boldsymbol{p}_1 \\ \boldsymbol{p}_2 \\ \vdots \\ \boldsymbol{p}_k \end{bmatrix} \tag{2.149}$$

and the matrix:

$$\boldsymbol{H} = \begin{bmatrix} \boldsymbol{H}_1 & \boldsymbol{0} & \cdots & \boldsymbol{0} \\ \boldsymbol{0} & \boldsymbol{H}_2 & \cdots & \boldsymbol{0} \\ \vdots & \vdots & \ddots & \vdots \\ \boldsymbol{0} & \boldsymbol{0} & \cdots & \boldsymbol{H}_k \end{bmatrix} \tag{2.150}$$

From Eq. (2.148) it's easy to verify that:

$$\boldsymbol{V}_0 = \boldsymbol{\Phi}_l \boldsymbol{H} \boldsymbol{p} \tag{2.151}$$

where:

$$\boldsymbol{\Phi}_l = \begin{bmatrix} \boldsymbol{I} & \boldsymbol{0} & \cdots & \boldsymbol{0} \\ {}^2_1\boldsymbol{\theta} & \boldsymbol{I} & \cdots & \boldsymbol{0} \\ \vdots & \vdots & \ddots & \boldsymbol{0} \\ {}^n_1\boldsymbol{\theta} & {}^n_2\boldsymbol{\theta} & \cdots & \boldsymbol{I} \end{bmatrix}, \quad {}^i_j\boldsymbol{\theta} = {}_{i-1}^{\;\;i}\boldsymbol{\theta} \ldots {}^{j+1}_{\;\;j}\boldsymbol{\theta}, \quad i = 2 \ldots n, \quad j = 1 \ldots n-1 \tag{2.152}$$

The dimension of the matrix $\boldsymbol{\Phi}_l \boldsymbol{H}$ in Eq. (2.151) is $6n \times n$. It can be partitioned into k blocks such as:

$$\boldsymbol{V}_0 = \begin{bmatrix} {}^1\dot{\boldsymbol{X}}_{1/0} \\ {}^2\dot{\boldsymbol{X}}_{2/0} \\ \vdots \\ {}^k\dot{\boldsymbol{X}}_{k/0} \end{bmatrix} = \begin{bmatrix} {}^1\boldsymbol{J}_{1/0}(O_1) \\ {}^2\boldsymbol{J}_{2/0}(O_2) \\ \vdots \\ {}^k\boldsymbol{J}_{k/0}(O_k) \end{bmatrix} \boldsymbol{p} \tag{2.153}$$

where:

$${}^i\dot{\boldsymbol{X}}_{P/0} = {}^i\boldsymbol{J}_{j/0}\left({}^jP\right)\boldsymbol{p} \tag{2.154}$$

The matrix ${}^i\boldsymbol{J}_{j/0}\left({}^jP\right)$ is known as the *Jacobian* of the structure. Similarly, the matrix $\boldsymbol{\Phi}_l \boldsymbol{H}$ of Eq. (2.151) can be regarded as the *global Jacobian* of the robot.

2.2.3.1 Extension to Robotic Chains with Ramified Topology

The matrix $\boldsymbol{\Phi}_l$ of Eq. (2.152) has been computed in case of linear chains. Let's now generalize its computation in presence of a kinematic chain with a ramified topology.

Let's consider a generic structure described as in Sect. 2.1.3 and having B branches, with each branch i made of a linear sequence of k_i links. Each link of the structure can be identified with the index pair (i, j), with $1 \leq i \leq B$ and $1 \leq j \leq k_i$. The joint associated with the aforementioned link (i, j) will have the following specifications:

- $r_{i,j}$ degrees of freedom
- joint matrix $\boldsymbol{H}_{i,j} \in \Re^{6 \times r_{i,j}}$
- configuration vector $\boldsymbol{q}_{i,j} \in \Re^{r_{i,j}}$
- quasivelocity vector $\boldsymbol{p}_{i,j} \in \Re^{r_{i,j}}$

The *global configuration vector* is then defined as:

$$\boldsymbol{q} \triangleq \left[\boldsymbol{q}_{1,1}{}^T \cdots \boldsymbol{q}_{1,k_1}{}^T \; \boldsymbol{q}_{2,1}{}^T \cdots \boldsymbol{q}_{2,k_2}{}^T \cdots \boldsymbol{q}_{B,1}{}^T \cdots \boldsymbol{q}_{B,k_B}{}^T\right]^T \tag{2.155}$$

Similarly, the *global quasivelocity vector* of the structure is defined as:

$$\boldsymbol{p} \triangleq \left[\boldsymbol{p}_{1,1}{}^T \cdots \boldsymbol{p}_{1,k_1}{}^T \; \boldsymbol{p}_{2,1}{}^T \cdots \boldsymbol{p}_{2,k_2}{}^T \cdots \boldsymbol{p}_{B,1}{}^T \cdots \boldsymbol{p}_{B,k_B}{}^T\right]^T \tag{2.156}$$

Finally, after defining the two matrices $\boldsymbol{H}$ and $\boldsymbol{V}_0$ as:

$$\boldsymbol{H} \triangleq diag\left[\boldsymbol{H}_{1,1}{}^T \cdots \boldsymbol{H}_{1,k_1}{}^T \; \boldsymbol{H}_{2,1}{}^T \cdots \boldsymbol{H}_{2,k_2}{}^T \cdots \boldsymbol{H}_{B,1}{}^T \cdots \boldsymbol{H}_{B,k_B}{}^T\right]^T \tag{2.157}$$

$$V_0 \triangleq \begin{bmatrix} {}^{(1,1)}\dot{X}_{(1,1)/0} \\ \vdots \\ {}^{(1,k_1)}\dot{X}_{(1,k_1)/0} \\ {}^{(2,1)}\dot{X}_{(2,1)/0} \\ \vdots \\ {}^{(2,k_2)}\dot{X}_{(2,k_2)/0} \\ \vdots \\ {}^{(B,1)}\dot{X}_{(B,1)/0} \\ \vdots \\ {}^{(B,k_B)}\dot{X}_{(B,k_B)/0} \end{bmatrix} \tag{2.158}$$

our goal is to find a representation of the global generalized velocity vector $\boldsymbol{v}_0$ with a relationship similar to Eq. (2.151), that is:

$$V_0 = \Phi_l H p \tag{2.159}$$

The recursive formula (2.146) is still valid, provided that it will be evaluated on the unique path that, beginning from the link in consideration, ends to the base frame.

To better understand the recursive process, lets compute the generalized velocity of the frame $\langle 2, 3\rangle$ of Fig. 2.16. We obtain the following recursion:

$$\begin{aligned}
{}^{(1,1)}\dot{X}_{(1,1)/(1,0)} &= H_{(1,1)}p_{(1,1)} \\
{}^{(1,2)}\dot{X}_{(1,2)/(1,0)} &= \begin{bmatrix} {}^{(1,2)}_{(1,1)}R & 0 \\ 0 & {}^{(1,2)}_{(1,1)}R \end{bmatrix} \phi\left({}^{(1,1)}r^{oj}_{(1,1)}\right) {}^{(1,1)}\dot{X}_{(1,1)/(1,0)} + H_{(1,2)}p_{(1,2)} \\
{}^{(1,3)}\dot{X}_{(1,3)/(1,0)} &= \begin{bmatrix} {}^{(1,3)}_{(1,2)}R & 0 \\ 0 & {}^{(1,3)}_{(1,2)}R \end{bmatrix} \phi\left({}^{(2,2)}r^{oj}_{(2,2)}\right) {}^{(1,2)}\dot{X}_{(1,2)/(1,0)} + H_{(1,3)}p_{(1,3)} \\
{}^{(2,1)}\dot{X}_{(2,1)/(1,0)} &= \begin{bmatrix} {}^{(2,1)}_{(1,3)}R & 0 \\ 0 & {}^{(2,1)}_{(1,3)}R \end{bmatrix} \phi\left({}^{(1,3)}r^{oj}_{(1,3)}\right) {}^{(1,3)}\dot{X}_{(1,3)/(1,0)} + H_{(2,1)}p_{(2,1)} \\
{}^{(2,2)}\dot{X}_{(2,2)/(1,0)} &= \begin{bmatrix} {}^{(2,2)}_{(2,1)}R & 0 \\ 0 & {}^{(2,2)}_{(2,1)}R \end{bmatrix} \phi\left({}^{(2,1)}r^{oj}_{(2,1)}\right) {}^{(2,1)}\dot{X}_{(2,1)/(1,0)} + H_{(2,2)}p_{(2,2)} \\
{}^{(2,3)}\dot{X}_{(2,3)/(1,0)} &= \begin{bmatrix} {}^{(2,3)}_{(2,2)}R & 0 \\ 0 & {}^{(2,3)}_{(2,2)}R \end{bmatrix} \phi\left({}^{(2,2)}r^{oj}_{(2,2)}\right) {}^{(2,2)}\dot{X}_{(2,2)/(1,0)} + H_{(2,3)}p_{(2,3)}
\end{aligned} \tag{2.160}$$

Similarly to the case of one linear chain, it is convenient to express the whole process in a global form. Let the matrix Φ be the following:

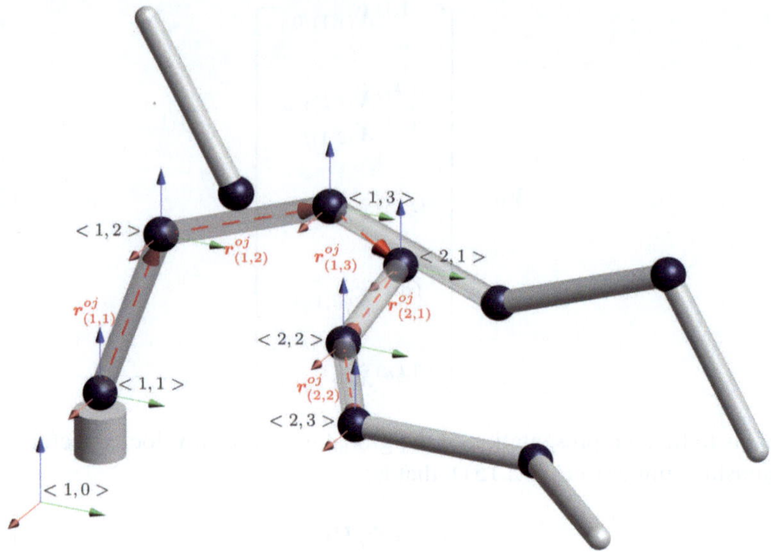

Fig. 2.16 Jacobian in a ramified chain: computation example

$$
\boldsymbol{\Phi} = \left[\begin{array}{cccc|cccc|c|cccc}
\boldsymbol{I} & \mathbf{0} & \cdots & \mathbf{0} & \mathbf{0} & \mathbf{0} & \cdots & \mathbf{0} & \cdots & \mathbf{0} & \mathbf{0} & \cdots & \mathbf{0} \\
{}^{(1,2)}_{(1,1)}\boldsymbol{\psi} & \boldsymbol{I} & \cdots & \mathbf{0} & \mathbf{0} & \mathbf{0} & \cdots & \mathbf{0} & \cdots & \mathbf{0} & \mathbf{0} & \cdots & \mathbf{0} \\
\vdots & \vdots & \ddots & \vdots & \vdots & \vdots & \ddots & \vdots & & \vdots & \vdots & \ddots & \vdots \\
{}^{(1,k_1)}_{(1,1)}\boldsymbol{\psi} & {}^{(1,k_1)}_{(1,2)}\boldsymbol{\psi} & \cdots & \boldsymbol{I} & \mathbf{0} & \mathbf{0} & \cdots & \mathbf{0} & \cdots & \mathbf{0} & \mathbf{0} & \cdots & \mathbf{0} \\
\hline
{}^{(2,1)}_{(1,1)}\boldsymbol{\psi} & {}^{(2,1)}_{(1,2)}\boldsymbol{\psi} & \cdots & {}^{(2,1)}_{(1,k_1)}\boldsymbol{\psi} & \boldsymbol{I} & \mathbf{0} & \cdots & \mathbf{0} & \cdots & \mathbf{0} & \mathbf{0} & \cdots & \mathbf{0} \\
{}^{(2,2)}_{(1,1)}\boldsymbol{\psi} & {}^{(2,2)}_{(1,2)}\boldsymbol{\psi} & \cdots & {}^{(2,2)}_{(1,k_1)}\boldsymbol{\psi} & {}^{(2,2)}_{(2,1)}\boldsymbol{\psi} & \boldsymbol{I} & \cdots & \mathbf{0} & \cdots & \mathbf{0} & \mathbf{0} & \cdots & \mathbf{0} \\
\vdots & \vdots & \ddots & \vdots & \vdots & \vdots & \ddots & \vdots & & \vdots & \vdots & \ddots & \vdots \\
{}^{(2,k_2)}_{(1,1)}\boldsymbol{\psi} & {}^{(2,k_2)}_{(1,2)}\boldsymbol{\psi} & \cdots & {}^{(2,k_2)}_{(1,k_1)}\boldsymbol{\psi} & {}^{(2,k_2)}_{(2,1)}\boldsymbol{\psi} & {}^{(2,k_2)}_{(2,2)}\boldsymbol{\psi} & \cdots & \boldsymbol{I} & \cdots & \mathbf{0} & \mathbf{0} & \cdots & \mathbf{0} \\
\hline
\vdots & \vdots & & \vdots & \vdots & \vdots & & \vdots & \ddots & \vdots & \vdots & & \vdots \\
\hline
{}^{(B,1)}_{(1,1)}\boldsymbol{\psi} & {}^{(B,1)}_{(1,2)}\boldsymbol{\psi} & \cdots & {}^{(B,1)}_{(1,k_1)}\boldsymbol{\psi} & {}^{(B,1)}_{(2,1)}\boldsymbol{\psi} & {}^{(B,1)}_{(2,2)}\boldsymbol{\psi} & \cdots & {}^{(B,1)}_{(2,k_2)}\boldsymbol{\psi} & \cdots & \boldsymbol{I} & \mathbf{0} & \cdots & \mathbf{0} \\
{}^{(B,2)}_{(1,1)}\boldsymbol{\psi} & {}^{(B,2)}_{(1,2)}\boldsymbol{\psi} & \cdots & {}^{(B,2)}_{(1,k_1)}\boldsymbol{\psi} & {}^{(B,2)}_{(2,1)}\boldsymbol{\psi} & {}^{(B,2)}_{(2,2)}\boldsymbol{\psi} & \cdots & {}^{(B,2)}_{(2,k_2)}\boldsymbol{\psi} & \cdots & {}^{(B,2)}_{(B,1)}\boldsymbol{\psi} & \boldsymbol{I} & \cdots & \mathbf{0} \\
\vdots & \vdots & \ddots & \vdots & \vdots & \vdots & \ddots & \vdots & & \vdots & \vdots & \ddots & \vdots \\
{}^{(B,k_B)}_{(1,1)}\boldsymbol{\psi} & {}^{(B,k_B)}_{(1,2)}\boldsymbol{\psi} & \cdots & {}^{(B,k_B)}_{(1,k_1)}\boldsymbol{\psi} & {}^{(B,k_B)}_{(2,1)}\boldsymbol{\psi} & {}^{(B,k_B)}_{(2,2)}\boldsymbol{\psi} & \cdots & {}^{(B,k_B)}_{(2,k_2)}\boldsymbol{\psi} & \cdots & {}^{(B,k_B)}_{(B,1)}\boldsymbol{\psi} & {}^{(B,k_B)}_{(B,2)}\boldsymbol{\psi} & \cdots & \boldsymbol{I}
\end{array}\right] \tag{2.161}
$$

Despite the apparent complexity due to a massive presence of indexes, the matrix ${}^{(r,s)}_{(i,j)}\boldsymbol{\psi}$ can be obtained through a simple recursive algorithm, obtained extrapolating the previous example. We have:

$$
{}^{(r,s)}_{(i,j)}\boldsymbol{\Psi} = \begin{cases}
\boldsymbol{0}^{6\times 6} & r < i \\
\boldsymbol{0}^{6\times 6} & s < j \ \ r = i \\
\boldsymbol{I}^{6\times 6} & s = j \ \ r = i \\
\begin{bmatrix} {}^{(r,s)}_{(r,s-1)}\boldsymbol{R} & \boldsymbol{0} \\ \boldsymbol{0} & {}^{(r,s)}_{(r,s-1)}\boldsymbol{R} \end{bmatrix} \boldsymbol{\phi}\left({}^{(r,s-1)}\boldsymbol{r}^{oj}_{(r,s-1)}\right) {}^{(r,s)}_{(i,s-1)}\boldsymbol{\Psi} & s > j \ \ r = i \\
\begin{bmatrix} {}^{(r,1)}_{(B_r,L_r)}\boldsymbol{R} & \boldsymbol{0} \\ \boldsymbol{0} & {}^{(r,1)}_{(B_r,L_r)}\boldsymbol{R} \end{bmatrix} \boldsymbol{\phi}\left({}^{(B_r,L_r)}\boldsymbol{r}^{B_r}_{(B_r,L_r)}\right) {}^{(r,1)}_{(B_r,L_r)}\boldsymbol{\Psi} & s = 1 \ \ r > i \\
\begin{bmatrix} {}^{(r,s)}_{(r,s-1)}\boldsymbol{R} & \boldsymbol{0} \\ \boldsymbol{0} & {}^{(r,s)}_{(r,s-1)}\boldsymbol{R} \end{bmatrix} \boldsymbol{\phi}\left({}^{(r,s-1)}\boldsymbol{r}^{oj}_{(r,s-1)}\right) {}^{(r,s)}_{(r,s-1)}\boldsymbol{\Psi} & s = 1 \ \ r > i
\end{cases}
\tag{2.162}
$$

where, as usual, ${}^{(r,s-1)}\boldsymbol{r}^{oj}_{(r,s-1)}$ is the free vector indicating the connection point of the outboard joint s of the branch r, expressed in the frame $\langle r,\ s-1\rangle$, while ${}^{(B_r,L_r)}\boldsymbol{r}^{B_r}_{(B_r,L_r)}$ indicates the connection point of the outboard joint 1 of the branch r, expressed in the frame $\langle B_r,\ L_r\rangle$ (the branch r is connected to the link L_r, belonging to the branch B_r).

With this formulation Eq. (2.159) holds and, similarly to the case of linear chains, the product $\boldsymbol{\Phi H}$ is the *global Jacobian* of the robot. It can be partitioned into k blocks such as:

$$
\boldsymbol{v}_0 = \begin{bmatrix}
{}^{(1,1)}\dot{\boldsymbol{X}}_{(1,1)/(1,0)} \\ \vdots \\ {}^{(1,k_1)}\dot{\boldsymbol{X}}_{(1,k_1)/(1,0)} \\ {}^{(2,1)}\dot{\boldsymbol{X}}_{(2,1)/(1,0)} \\ \vdots \\ {}^{(2,k_2)}\dot{\boldsymbol{X}}_{(2,k_2)/(1,0)} \\ \vdots \\ {}^{(B,1)}\dot{\boldsymbol{X}}_{(B,1)/(1,0)} \\ \vdots \\ {}^{(B,k_B)}\dot{\boldsymbol{X}}_{(B,k_B)/(1,0)}
\end{bmatrix} = \begin{bmatrix}
{}^{(1,1)}\boldsymbol{J}_{(1,1)/(1,0)} \\ \vdots \\ {}^{(1,k_1)}\boldsymbol{J}_{(1,k_1)/(1,0)} \\ {}^{(2,1)}\boldsymbol{J}_{(2,1)/(1,0)} \\ \vdots \\ {}^{(2,k_2)}\boldsymbol{J}_{(2,k_2)/(1,0)} \\ \vdots \\ {}^{(B,1)}\boldsymbol{J}_{(B,1)/(1,0)} \\ \vdots \\ {}^{(B,k_B)}\boldsymbol{J}_{(B,k_B)/(1,0)}
\end{bmatrix} \boldsymbol{p}
\tag{2.163}
$$

where the matrix ${}^{(r,s)}\boldsymbol{J}_{(i,j)/(1,0)}\left({}^{(i,j)}P\right)$ is the *Jacobian* of the structure.

In other words, given a generic point ${}^{(i,j)}P$ integral to the frame $\langle i,\ j\rangle$, the Jacobian transforms the global quasivelocity vector into the generalized velocity of the point P:

$$
{}^{(r,s)}\dot{\boldsymbol{X}}_{P/(1,0)} = {}^{(r,s)}\boldsymbol{J}_{(i,j)/(1,0)}\left({}^{(i,j)}P\right)\boldsymbol{p}. \tag{2.164}
$$

2.3 Dynamics

This section faces the problem of searching a mathematical model that fully describes the dynamics of the manipulation system, made of open (and possibly organized with a ramified topology) chains.

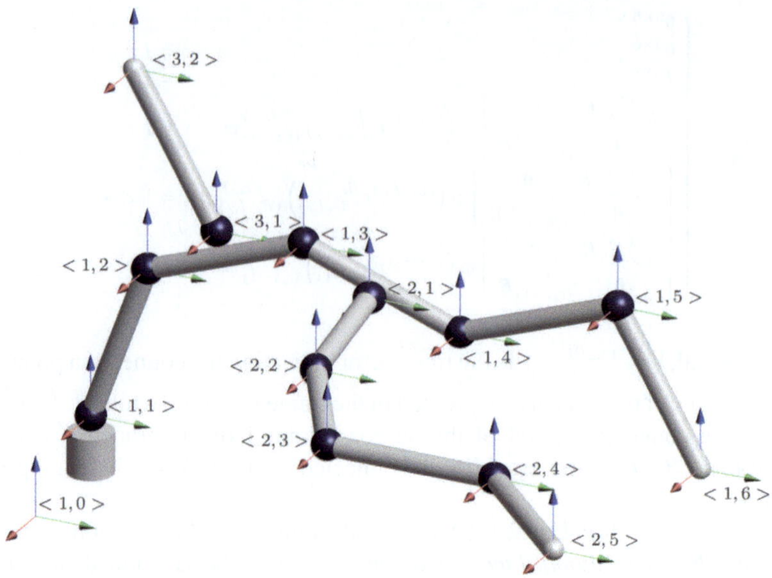

Fig. 2.17 Open kinematic chain with multiple branch topology

Similarly to many solutions proposed in literature [3, 6], the study is based on the Lagrange equations. We'll compute the Lagrange equations in terms of the robot descriptive language introduced in the previous sections and, after some algebraic manipulations, they will be reduced to the typical differential equations system that describes the behavior of a mechanical system.

2.3.1 Equilibrium of a Manipulation Structure

As described in Sect. 2.1.3, we will consider only open chains with ramified topology. Figure 2.17 shows an example of such a multi-body system. For each rigid body constituting the structure, the equilibrium conditions are [6]:

$$\begin{cases} \boldsymbol{R}_i = \boldsymbol{0} \\ \boldsymbol{M}_i = \boldsymbol{0} \end{cases} \tag{2.165}$$

where $\boldsymbol{R}_i$ is the resultant of all the external forces acting on the body and $\boldsymbol{M}_i$ is the resultant of all the external momentums.

By applying the virtual work principle [6] we obtain:

$$\begin{cases} \boldsymbol{R}_i \cdot \partial P_i = \boldsymbol{0} \\ \boldsymbol{M}_i \cdot \partial \theta_i = \boldsymbol{0} \end{cases} \tag{2.166}$$

being P_i the point of application of $\boldsymbol{R}_i$. Extending to the whole structure we have:

$$\sum_{b=1}^{B}\sum_{i=1}^{k_b}\boldsymbol{R}_{(b,i)}\cdot\partial P_{(b,i)}+\sum_{b=1}^{B}\sum_{i=1}^{k_b}\boldsymbol{M}_{(b,i)}\cdot\partial\theta_{(b,i)}=0 \tag{2.167}$$

where B is the number of branches of the structure and k_b is the number of links of the branch b.

Considering the joints ideally smooth, the virtual work produced by the internal joint reaction is globally null and Eq. (2.166) becomes:

$$\sum_{b=1}^{B}\sum_{i=1}^{k_b}\boldsymbol{R}^{(a)}_{(b,i)}\cdot\partial P_{(b,i)}+\sum_{b=1}^{B}\sum_{i=1}^{k_b}\boldsymbol{M}^{(a)}_{(b,i)}\cdot\partial\theta_{(b,i)}=\boldsymbol{0} \tag{2.168}$$

where $\boldsymbol{R}^{(a)}_{(b,i)}$ and $\boldsymbol{M}^{(a)}_{(b,i)}$ are the resultants of the active forces and torques.

The left-hand side of Eq. (2.168) represents the virtual work of the actuation actions $\boldsymbol{m}$ and of the external forces. After some algebraic manipulations we obtain:

$$\boldsymbol{m}+{}^{(1,0)}\boldsymbol{J}^{T}_{(h,k)/(1,0)}\begin{bmatrix}{}^{(1,0)}\boldsymbol{M}_{(h,k)}\\ {}^{(1,0)}\boldsymbol{R}_{(h,k)}\end{bmatrix}=\boldsymbol{0} \tag{2.169}$$

where the expression ${}^{(1,0)}\boldsymbol{J}^{T}_{(h,k)/(1,0)}\begin{bmatrix}{}^{(1,0)}\boldsymbol{M}_{(h,k)}\\ {}^{(1,0)}\boldsymbol{R}_{(h,k)}\end{bmatrix}$ represents the projection of the external forces in the joint space.

2.3.2 The Lagrange Equation

In its typical form, the Lagrange equation of motion is given by [3, 6]:

$$\frac{d}{dt}\frac{\partial T\left(\boldsymbol{q},\dot{\boldsymbol{q}}\right)}{\partial\dot{\boldsymbol{q}}}-\frac{\partial T\left(\boldsymbol{q},\dot{\boldsymbol{q}}\right)}{\partial\boldsymbol{q}}=\boldsymbol{\mu} \tag{2.170}$$

where:

- $\boldsymbol{q}$ global configuration vector (joint coordinates)
- $T\left(\boldsymbol{q},\dot{\boldsymbol{q}}\right)$ kinetic energy of the structure
- $\boldsymbol{\mu}$ projection of the external actions to the joint space

Considering that the joints may have more than one degree of freedom (as for example in the case of spherical joints), it may be convenient to formulate the motion equations in terms of variable different from the derivative of the configuration vector $\dot{\boldsymbol{q}}$. Those variables cannot be regarded as any exact differential and, as introduced in Sect. 2.2.2

(see Eq. 2.90), they are often known as *quasi-velocities* [6]. They may represent, for example, the angular velocities of the bodies constituting the links.

Objective of the following sections is to express the kinetic energy in those terms.

2.3.2.1 Kinetic Energy

In general, regardless of the chosen description, the kinetic energy is the sum of the energies of the links constituting the manipulator:

$$T = \frac{1}{2}\sum_{b=1}^{B}\sum_{i=1}^{k_b} m_{(b,i)}\,{}^{(k,j)}\boldsymbol{v}_{c(b,i)}^{T}\,{}^{(k,j)}\boldsymbol{v}_{c(b,i)} + {}^{(k,j)}\boldsymbol{\omega}_{(b,i)}^{T}\left[{}^{(k,j)}_{(b,i)}\boldsymbol{R}\,{}^{(b,i)}\boldsymbol{I}_{(b,i)}\,{}^{(k,j)}_{(b,i)}\boldsymbol{R}^{T}\right]{}^{(k,j)}\boldsymbol{\omega}_{(b,i)} \tag{2.171}$$

where:

- B number of branches
- k_b number of links of the branch b
- $m_{(b,i)}$ mass of the link i of the branch b
- ${}^{(k,j)}\boldsymbol{\omega}_{(b,i)}$ angular velocity of the frame $\langle b, i\rangle$ projected on $\langle k, j\rangle$
- ${}^{(k,j)}\boldsymbol{v}_{c(b,i)}$ cartesian velocity of the center of mass of the link (b, i) projected on the frame $\langle k, j\rangle$
- ${}^{(k,j)}_{(b,i)}\boldsymbol{R}$ rotation matrix of the frame $\langle b, i\rangle$ w.r.t $\langle k, j\rangle$
- ${}^{(b,i)}\boldsymbol{I}_{(b,i)}$ inertia tensor of the link (b, i) computed w.r.t. a frame parallel to $\langle b, i\rangle$ and with the origin placed in the center of mass $c_{(b,i)}$ of the same link

Being the terms of the sum all scalars, all the quantities in Eq. (2.171) can be projected on any frame (per given sum element). Here, the choice is to evaluate all the velocities on the same frame to which they belong, with the advantage of a simplified recursive algorithm and the elimination of the transformation of the inertia tensor.

Let's define a *spatial inertia tensor* as:

$$\boldsymbol{M}_{(b,i)} = \begin{bmatrix} {}^{(b,i)}\boldsymbol{I}_{(b,i)} & \boldsymbol{0} \\ \boldsymbol{0} & m_{(b,i)}\boldsymbol{I}_3 \end{bmatrix} \tag{2.172}$$

with $\boldsymbol{I}_3$ being the 3×3 identity matrix. Assuming $(k, j) = (b, i)$, Eq. (2.171) becomes:

$$\begin{aligned} T &= \frac{1}{2}\sum_{b=1}^{B}\sum_{i=1}^{k_b} m_{(b,i)}\,{}^{(b,i)}\boldsymbol{v}_{c(b,i)}^{T}\,{}^{(b,i)}\boldsymbol{v}_{c(b,i)} + {}^{(b,i)}\boldsymbol{\omega}_{(b,i)}^{T}\,{}^{(b,i)}\boldsymbol{I}_{(b,i)}\,{}^{(b,i)}\boldsymbol{\omega}_{(b,i)} \\ &= \frac{1}{2}\sum_{b=1}^{B}\sum_{i=1}^{k_b} {}^{(b,i)}\dot{\boldsymbol{X}}_{c(b,i)}^{T}\boldsymbol{M}_{(b,i)}\,{}^{(b,i)}\dot{\boldsymbol{X}}_{c(b,i)} \\ &= \boldsymbol{V}_{cm}^{T}\boldsymbol{M}_{cm}\boldsymbol{V}_{cm} \end{aligned} \tag{2.173}$$

where:

$$V_{cm} = \begin{bmatrix} {}^{(1,1)}\dot{X}_{c_{(1,1)}/(1,0)} \\ \vdots \\ {}^{(1,k_1)}\dot{X}_{c_{(1,k_1)}/(1,0)} \\ {}^{(2,1)}\dot{X}_{c_{(2,1)}/(1,0)} \\ \vdots \\ {}^{(2,k_2)}\dot{X}_{c_{(2,k_2)}/(1,0)} \\ \vdots \\ {}^{(B,1)}\dot{X}_{c_{(B,1)}/(1,0)} \\ \vdots \\ {}^{(B,1)}\dot{X}_{c_{(B,k_B)}/(1,0)} \end{bmatrix} \tag{2.174}$$

$$M_{cm} = diag\left[M_{(1,1)} \cdots M_{(1,k)}\ M_{(2,1)} \cdots M_{(2,k_2)} \cdots M_{(B,1)} \cdots M_{(B,k_B)} \right] \tag{2.175}$$

Recalling Eqs. (2.141) and (2.142), the global generalized velocity of the center of mass (2.174) becomes:

$$V_{cm} = \begin{bmatrix} \phi\left({}^{(1,1)}r^{cm}_{(1,1)}\right) & \cdots & 0 \\ \vdots & \ddots & \vdots \\ 0 & \cdots & \phi\left({}^{(B,k_B)}r^{cm}_{(B,k_b)}\right) \end{bmatrix} v_0 \tag{2.176}$$

with V_0 defined from Eq. (2.158). Substituting Eq. (2.176) into (2.173) we obtain:

$$T = \frac{1}{2}V_0^T \begin{bmatrix} \phi^*\left({}^{(1,1)}r^{cm}_{(1,1)}\right) & \cdots & 0 \\ \vdots & \ddots & \vdots \\ 0 & \cdots & \phi^*\left({}^{(B,k_B)}r^{cm}_{(B,k_B)}\right) \end{bmatrix} M_{cm} \begin{bmatrix} \phi\left({}^{(1,1)}r^{cm}_{(1,1)}\right) & \cdots & 0 \\ \vdots & \ddots & \vdots \\ 0 & \cdots & \phi\left({}^{(B,k_B)}r^{cm}_{(B,k_B)}\right) \end{bmatrix} v_0$$

$$= \frac{1}{2}p^T H^T \Phi^T \begin{bmatrix} \phi^*\left({}^{(1,1)}r^{cm}_{(1,1)}\right) & \cdots & 0 \\ \vdots & \ddots & \vdots \\ 0 & \cdots & \phi^*\left({}^{(B,k_B)}r^{cm}_{(B,k_B)}\right) \end{bmatrix} M_{cm} \begin{bmatrix} \phi\left({}^{(1,1)}r^{cm}_{(1,1)}\right) & \cdots & 0 \\ \vdots & \ddots & \vdots \\ 0 & \cdots & \phi\left({}^{(B,k_B)}r^{cm}_{(B,k_B)}\right) \end{bmatrix} \Phi H p$$

$$= \frac{1}{2}\mathbf{p}^T H^T \Phi^T M \Phi H \mathbf{p} = \frac{1}{2}\mathbf{p}^T A \mathbf{p} \tag{2.177}$$

Hence:

$$T = \frac{1}{2}p^T H^T \Phi^T M \Phi H p = \frac{1}{2}p^T A p \tag{2.178}$$

where:

$$\boldsymbol{M} = \begin{bmatrix} \phi^*\left({}^{(1,1)}\boldsymbol{r}^{cm}_{(1,1)}\right)\boldsymbol{M}_{(1,1)}\phi\left({}^{(1,1)}\boldsymbol{r}^{cm}_{(1,1)}\right) & \cdots & \boldsymbol{0} \\ \vdots & \ddots & \vdots \\ \boldsymbol{0} & \cdots & \phi^*\left({}^{(B,k_B)}\boldsymbol{r}^{cm}_{(B,k_B)}\right)\boldsymbol{M}_{(B,k_B)}\phi\left({}^{(B,k_B)}\boldsymbol{r}^{cm}_{(B,k_B)}\right) \end{bmatrix} \quad (2.179)$$

The matrix

$$\boldsymbol{A} = \boldsymbol{H}^T\boldsymbol{\Phi}^T\boldsymbol{M}\boldsymbol{\Phi}\boldsymbol{H} \quad (2.180)$$

is known as the *inertia matrix* of the structure.

We have hence expressed the kinetic energy T in terms of the vectors $\boldsymbol{q}$ and $\boldsymbol{p}$ in a compact matrix form. This form is particularly convenient when using symbolic processors, allowing also a certain level of optimization in case of a reasonable number of degrees of freedom.

2.3.2.2 The Lagrange Equation for Quasi-Coordinates

We have just expressed the kinetic energy as a function $\overline{T}(\boldsymbol{q},\boldsymbol{p})$ in place of $T(\boldsymbol{q},\dot{\boldsymbol{q}})$. The next step is to evaluate the consequent transformation of the Lagrange equation.

As already seen in Eq. (2.112), the relationship between $\boldsymbol{p}$ and $\dot{\boldsymbol{q}}$ has the form:

$$\dot{\boldsymbol{q}} = \boldsymbol{V}\boldsymbol{p} \quad (2.181)$$

or, equivalently:

$$\boldsymbol{p} = \boldsymbol{\alpha}\dot{\boldsymbol{q}} \quad (2.182)$$

with $\boldsymbol{\alpha} = \boldsymbol{V}^{-1}$. With this substitution the derivatives $\frac{\partial T}{\partial \dot{q}_k}$ becomes:

$$\frac{\partial T}{\partial \dot{q}_k} = \sum_{i=1}^{n} \frac{\partial \overline{T}}{\partial p_i}\frac{\partial p_i}{\partial \dot{q}_k} \quad (2.183)$$

Equation (2.170) then becomes:

$$\frac{d}{dt}\left\{\frac{\partial \overline{T}(\boldsymbol{p},\boldsymbol{q})}{\partial \boldsymbol{p}}\right\} + \boldsymbol{\alpha}^T[\boldsymbol{\beta}-\gamma]\left\{\frac{\partial \overline{T}(\boldsymbol{p},\boldsymbol{q})}{\partial \boldsymbol{p}}\right\} - \boldsymbol{\alpha}^T\left\{\frac{\partial \overline{T}(\boldsymbol{p},\boldsymbol{q})}{\partial \boldsymbol{q}}\right\} = \boldsymbol{V}^T\boldsymbol{\mu} \quad (2.184)$$

where:

$$\beta_{i,j} = \boldsymbol{p}^T\boldsymbol{\alpha}^T\left\{\frac{\partial\left(\boldsymbol{\alpha}^{-T}\right)_{i,j}}{\partial \boldsymbol{q}}\right\} \quad (2.185)$$

$$\gamma = \begin{bmatrix} p^T \alpha^T \frac{\partial(\alpha^{-T})}{\partial q_1} \\ p^T \alpha^T \frac{\partial(\alpha^{-T})}{\partial q_2} \\ \vdots \\ p^T \alpha^T \frac{\partial(\alpha^{-T})}{\partial q_n} \end{bmatrix} \tag{2.186}$$

From the expression of the kinetic energy of Eq. (2.178) we have:

$$\frac{\partial \overline{T}\,(\boldsymbol{p}, \boldsymbol{q})}{\partial \boldsymbol{p}} = \boldsymbol{A}\boldsymbol{p} \tag{2.187}$$

and:

$$\frac{d}{dt}\left\{\frac{\partial \overline{T}\,(\boldsymbol{p}, \boldsymbol{q})}{\partial \boldsymbol{p}}\right\} = \frac{d}{dt}\boldsymbol{A}\boldsymbol{p} = \boldsymbol{A}\dot{\boldsymbol{p}} + \dot{\boldsymbol{A}}\boldsymbol{p} = \boldsymbol{A}\dot{\boldsymbol{p}} + \left[\frac{\partial \boldsymbol{A}\boldsymbol{p}}{\partial \boldsymbol{q}}\right]\boldsymbol{\alpha}\boldsymbol{p} \tag{2.188}$$

$$\frac{\partial \overline{T}\,(\boldsymbol{p}, \boldsymbol{q})}{\partial \boldsymbol{q}} = \frac{1}{2}\left[\frac{\partial \boldsymbol{A}\boldsymbol{p}}{\partial \boldsymbol{q}}\right]^T \boldsymbol{p} \tag{2.189}$$

Finally, Eq. (2.184) assumes the form:

$$\boldsymbol{A}\,(\boldsymbol{q})\dot{\boldsymbol{p}} + \boldsymbol{B}\,(\boldsymbol{q}, \boldsymbol{p})\boldsymbol{p} + \boldsymbol{V}^T \boldsymbol{C}\,(\boldsymbol{q}) = \boldsymbol{V}^T \boldsymbol{\mu}^e \tag{2.190}$$

where:

$$\boldsymbol{B}\,(\boldsymbol{q}, \boldsymbol{p}) = \left[\frac{\partial \boldsymbol{A}\boldsymbol{p}}{\partial \boldsymbol{q}}\right]\boldsymbol{V}^{-T} - \frac{1}{2}\boldsymbol{V}^{-T}\left[\frac{\partial \boldsymbol{A}\boldsymbol{p}}{\partial \boldsymbol{q}}\right]^T + \boldsymbol{V}^{-T}\,[\boldsymbol{\beta} - \boldsymbol{\gamma}]\,\boldsymbol{A} \tag{2.191}$$

Equation (2.190) is known as the *Lagrange equation for quasi-coordinates* (see also Meirovitch [6]).

Note that the quantity $\boldsymbol{V}^T \boldsymbol{C}\,(\boldsymbol{q})$ represents the projection in the joint velocity space of the gravitational forces, while $\boldsymbol{V}^T \boldsymbol{\mu}^e$ is the projection in the joint velocity space of all the other external forces (for example the actuator forces).

2.3.2.3 Example: Spherical Joint

Let's focus now on the computation of some quantities appearing within the Lagrange equation for quasi-coordinates (2.190) in case of a spherical joint. The latter can be seen as a one link system, connected to the main frame with a 3 DOF joint described by the joint matrix (2.98). Its parameterization in the variables $q1$, $q2$ and $q3$ has been already computed in Sect. 2.2.2.2.

The matrix $\boldsymbol{V}$ of Eq. (2.181) can be computed with the process introduced in Sect. 2.2.2.4, that gives the following result:

$$\boldsymbol{V} = \begin{bmatrix} 1 & \frac{\sin(q_1)\sin(q_2)}{\cos(q_2)} & \frac{\cos(q_1)\sin(q_2)}{\cos(q_2)} \\ 0 & \cos(q_1) & -\sin(q_1) \\ 0 & \frac{\sin(q_1)}{\cos(q_2)} & \frac{\cos(q_1)}{\cos(q_2)} \end{bmatrix} \tag{2.192}$$

where q_1, q_2 and q_3 are the joint variables (see Equations (2.100) through (2.106)). The computation of $\boldsymbol{B}(\boldsymbol{q}, \boldsymbol{p})$ by using Eq. (2.191) gives:

$$\boldsymbol{B}(\boldsymbol{q}, \boldsymbol{p}) = \boldsymbol{V}^{-T}[\beta - \gamma]\boldsymbol{A} = \begin{bmatrix} 0 & -p_3 & p_2 \\ p_3 & 0 & -p_1 \\ -p_2 & p_1 & 0 \end{bmatrix} \boldsymbol{A} \tag{2.193}$$

being $\boldsymbol{A}$ constant. Equation (2.193) shows the absence of discontinuities appearing within the matrix $\boldsymbol{V}$ of Eq. (2.192). This is one of the most noticeable advantages of the Lagrange equation for quasi-coordinates. In case of spherical joints parameterized with Euler angles, it holds also where the matrix $\boldsymbol{V}$ of Eq. (2.181) is non-invertible.

References

1. Denavit J, Hartenberg RS (1955) A kinematic notation for lower-pair mechanisms based on matrices. Trans ASME E J Appl Mech 22:215–221
2. McCarthy J (1990) An introduction to theoretical kinematics. MIT Press, Cambridge. http://books.google.com/books?id=glOqQgAACAAJ
3. Goldstein H (1980) Classical mechanics, 2nd edn. Addison-Wesley, Reading
4. Olver P (2000) Applications of lie groups to differential equations. Graduate texts in mathematics. Springer, Berlin. http://books.google.com/books?id=sI2bAxgLMXYC
5. Kwatny H, Blankenship G (1995) Symbolic construction of models for multibody dynamics. IEEE Trans Robot Autom 11(2):271–281. doi:10.1109/70.370509
6. Meirovitch L (1970) Methods of analytical dynamics. Advanced engineering series. McGraw-Hill, New York. http://books.google.com/books?id=6fNQAAAAMAAJ

Chapter 3
Kinematic Control

The primary purpose of an autonomous manipulation system is to perform intervention tasks with a limited exchange of information between the manipulator and the human supervisor. The information passed to the main control system is often only a high level decision command, and the controller must be capable of following the command by providing reliable control references to the actuators.

The main issue in designing and implementing a control system for autonomous manipulation is ensuring a reliable behavior within the workspace. A reliable behavior means also avoiding singularities, collisions, system instabilities and unwanted motions while performing the required task is theoretically executable.

The control system must also address some general manipulation issues, such as being task-space oriented, with task priority assignments and dynamic priority changes.

The third layer of the main control diagram of Fig. 3.1 is the Medium Level Controller of the system and it is the layer where the above issues are addressed. This Chapter describes some possible solutions to the inherent kinematical problems in a control system for autonomous manipulation. We chose a "task reconstruction" approach in order to automatically correct the required task according to the priority of the situation. For example, when approaching a singularity, in order to prevent unwanted motions and system instabilities the priority of the control system becomes maintaining the distance from the singularity over a predefined threshold, rather than following the input task. This approach is extendible to several other issues, such as collision avoidance, and ensures a reliable execution of the input task when it is theoretically feasible. If, for any reason, the input task cannot be executed, the control system must inform the above layer of the execution error, together with the nature of the problem (i.e. approaching a singularity).

G. Marani and J. Yuh, *Introduction to Autonomous Manipulation*,
Springer Tracts in Advanced Robotics 102, DOI: 10.1007/978-3-642-54613-6_3,

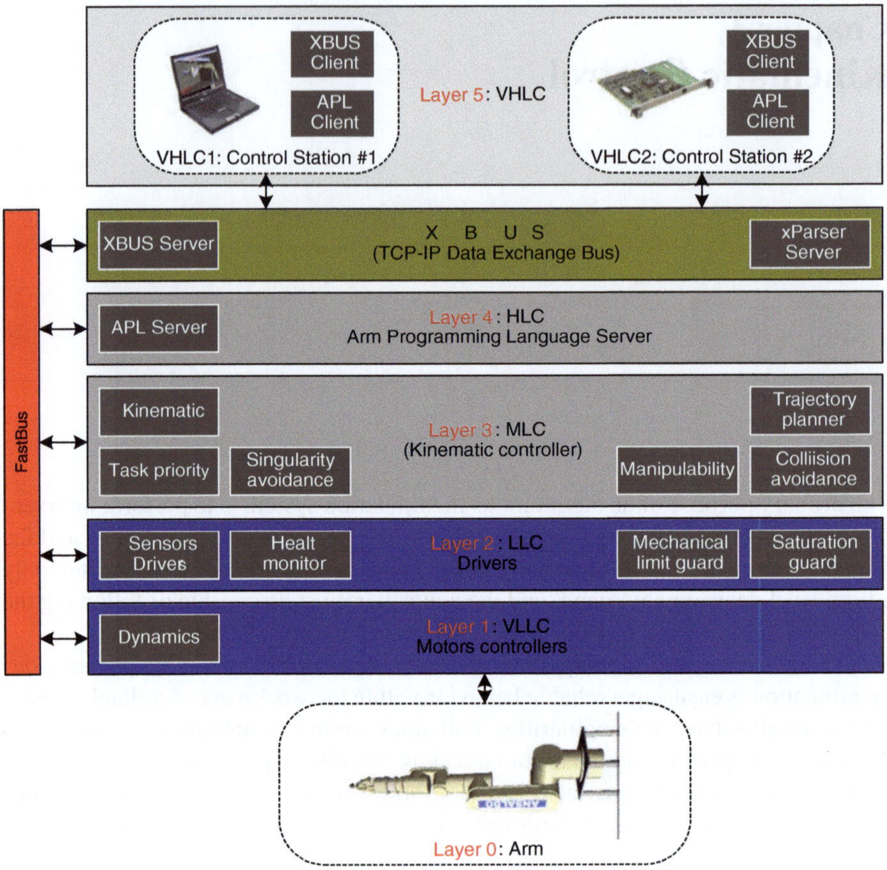

Fig. 3.1 Control system

3.1 Generation of the Velocity Reference

In this chapter we are not considering the inherent problem in the dynamic control of the manipulator. In our treatise, the arm together with its dynamic controller is considered as a separate subsystem, as shown in Fig. 3.2. The function of the kinematic controller described here is to generate an appropriate velocity profile for the actuators.

This approach is appropriate for a large class of electromechanical manipulators. As a matter of fact, the dynamic control is often performed by sophisticated hardware PID controllers, which are able to limit the tracking errors to acceptable values. This is especially true when the joints are decoupled from the motors by a high gear ratio that helps further reducing the joint velocity error.

In the control scheme of Fig. 3.2, the block named *Robot HW Controller* represents the physical manipulator equipped with its joint drives, each one implementing a

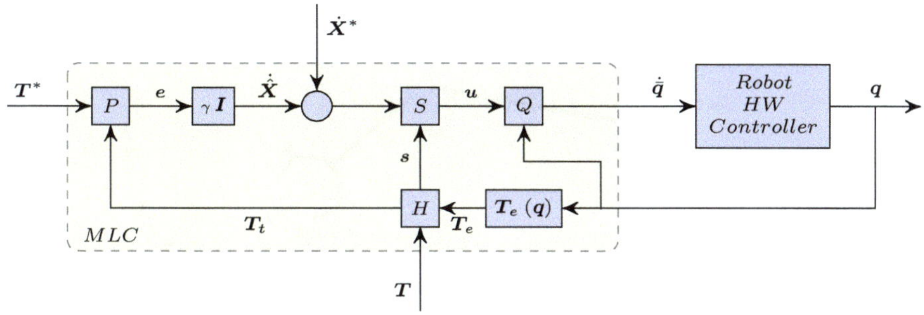

Fig. 3.2 Control scheme

closed loop velocity control at the corresponding joint. Using this control scheme, the robot can be regarded as a black-box, receiving the vector of the reference joint velocities as input, and giving the vector of the corresponding joint positions as output. The last substantially coincides with the time integral of $\boldsymbol{q}$, provided that sufficiently high bandwidth loops are guaranteed by the hardware controller itself.

For completeness, a detailed description of some motion control algorithms for the control scheme in Fig. 3.2 was presented in [1].

3.1.1 Closing the Feedback Loop

Let's consider a schematic representation of a generic manipulator and its workspace as shown in Fig. 3.3. Here, $\boldsymbol{T}_e$ is the transformation matrix of the end-effector frame $\langle e\rangle$ with respect to the base frame $\langle o\rangle$, $\boldsymbol{T}$ is the (constant) transformation matrix of the tool center frame $\langle t\rangle$ with respect to $\langle e\rangle$, while $\boldsymbol{T}^*$, generally time varying, is the transformation matrix of the reference frame $\langle g\rangle$ with respect to the base frame $\langle o\rangle$. The reference frame $\langle g\rangle$ is usually integral with the target, while the base frame $\langle o\rangle$ is integral with the vehicle. In the general case, both the above frames are time-dependent and moving with respect to the earth-fixed frame (not shown in Fig. 3.3).

The general goal is to track the reference frame $\langle g\rangle$ by the tool frame $\langle t\rangle$. At this aim, the global error e is automatically defined by the vector:

$$\boldsymbol{e} \doteq \left[\boldsymbol{r}_{gt}, \boldsymbol{\rho}_{gt}\right]^T \tag{3.1}$$

where vectors $\boldsymbol{r}_{gt}$ and $\boldsymbol{\rho}_{gt}$ (both projected on the base frame $\langle o\rangle$) represent the distance and the misalignment (equivalent rotation vector) of the reference frame $\langle g\rangle$ with respect to $\langle t\rangle$. The objective of the control scheme is to make the global error $\boldsymbol{e}$ asymptotically converging toward zero or, alternatively, asymptotically confined within acceptable norm bounds. This goal could be achieved with the closed loop scheme shown in Fig. 3.2.

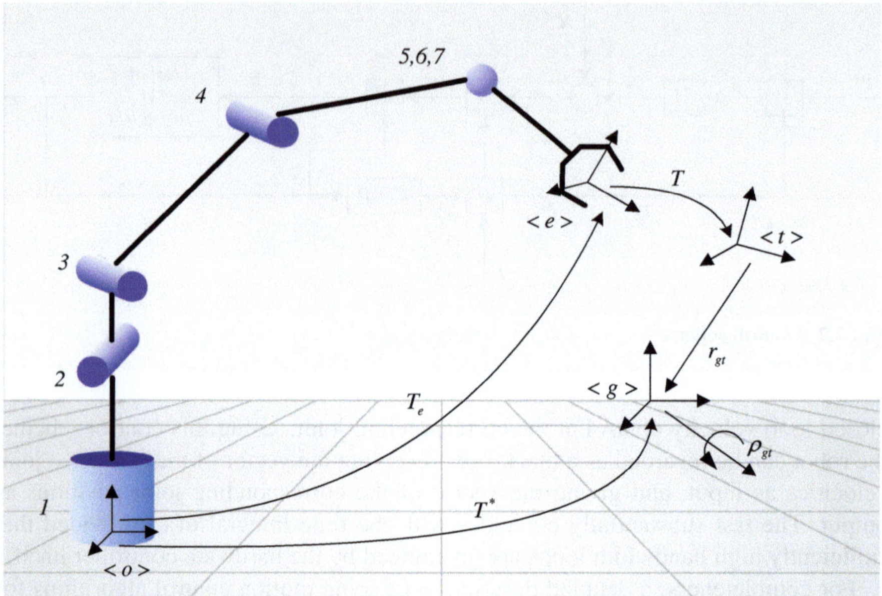

Fig. 3.3 Schematic representation of a 7 DOF robotic arm and its workspace

3.1.1.1 Medium Level Control Loop

The remaining part of the control system represents the Medium Level Control (MLC) loop of the arm. The joint velocity reference signals $\dot{\bar{q}}$ are appropriately generated as real-time outputs, such that the global error e converges toward the specified bounds. The reference transformation matrix T^* is compared with the actual tool frame transformation matrix $\boldsymbol{T}_t$ via the processing block P, which is used for evaluating the global error $\boldsymbol{e}$ in real time by solving, for the rotational error part $\boldsymbol{\rho}_{gt}$ only, the well known *Versor Lemma* equations [2], given by:

$$\begin{cases} \boldsymbol{i}_t \wedge \boldsymbol{i}^* + \boldsymbol{j}_t \wedge \boldsymbol{j}^* + \boldsymbol{k}_t \wedge \boldsymbol{k}^* = \frac{1}{2} z \sin\theta \\ \boldsymbol{i}_t^T \cdot \boldsymbol{i}^* + \boldsymbol{j}_t^T \cdot \boldsymbol{j}^* + \boldsymbol{k}_t^T \cdot \boldsymbol{k}^* = 1 + \cos\theta \end{cases} \tag{3.2}$$

where we assumed $\rho \doteq z\theta$, with z a unitary vector and θ an angular quantity, and where $R_t \doteq \left[\boldsymbol{i}_t, \boldsymbol{j}_t, \boldsymbol{k}_t\right]$ and $R^* \doteq \left[\boldsymbol{i}^*, \boldsymbol{j}^*, \boldsymbol{k}^*\right]$ are the rotation matrices contained inside the transformation matrices $\boldsymbol{T}_t$ and $\boldsymbol{T}^*$ respectively. The notation $\boldsymbol{a} \wedge \boldsymbol{b}$ is used for indicating the cross product of two generic three dimensional vectors $\boldsymbol{a}$ and $\boldsymbol{b}$. Proof of the Versor Lemma is given in Appendix A.1.

The linear part $\boldsymbol{r}_{gt}$ of the global error is easily obtained as difference between the first three elements of the last columns of $\boldsymbol{T}^*$ and those of $\boldsymbol{T}_t$. The global error $\boldsymbol{e}$ is then multiplied by a suitable gain matrix $\gamma\boldsymbol{I}$. The result is the generalized Cartesian velocity $\dot{\hat{\boldsymbol{X}}} \doteq \left[\hat{\omega}, \hat{\boldsymbol{v}}\right]^T \in \Re^6$ (projected on $\langle o \rangle$), where $\hat{\omega} \in \Re^3$ and $\hat{\boldsymbol{v}} \in \Re^3$ are the angular and linear velocity, respectively, which are assigned to the tool frame $\langle T_t \rangle$ such that $\boldsymbol{e}$ converges within the specified bounds. At this stage, the additional Cartesian velocity input $\dot{\boldsymbol{X}}^*$ allows a direct control of the end-effector velocity, which is needed for example for force-feedback control.

The generalized velocity control input $\dot{\hat{\boldsymbol{X}}}$ is then translated into the one-to-one related velocity $\dot{\bar{\boldsymbol{X}}}^{\cdot} = [\bar{\omega}, \bar{v}]$ to be assigned to the end-effector frame $\langle e \rangle$. The velocity translation is performed by the block S, using the well-known rigid body velocity relationships:

$$\begin{cases} \bar{\omega} = \hat{\omega} \\ \bar{v} = \hat{v} + [s \wedge]\, \hat{\omega} \end{cases} \tag{3.3}$$

where $\boldsymbol{s}$ is the vector distance (projected on $\langle o \rangle$) of the frame $\langle t \rangle$ with respect to $\langle e \rangle$, which is real time evaluated by the functional block H via the well known projection relationship:

$$\boldsymbol{s} = \boldsymbol{R}_e\,(\boldsymbol{q})\,\boldsymbol{t} \tag{3.4}$$

where $\boldsymbol{R}_e\,(\boldsymbol{q})$ is the rotation matrix part of the end effector frame transformation matrix $\boldsymbol{T}_e\,(\boldsymbol{q})$, while $\boldsymbol{t}$ is the same (constant) distance vector projected on the end effector frame $\langle e \rangle$ (i.e. the first three components of the last column of the constant transformation matrix $\boldsymbol{T}$).

The end-effector Cartesian velocity control signal $\dot{\bar{\boldsymbol{X}}}$ is transformed into a corresponding set of joint velocity reference input vector $\dot{\bar{\boldsymbol{q}}}$ by the functional block Q.

3.2 Inverse Kinematics

The interface block Q of the control scheme of Fig. 3.2 generates the velocity references for the joint actuators from a given task velocity reference input. Within this operation, addressed as kinematic inversion, we must consider all the previously mentioned issues such as task priority, singularities avoidance, joint limit and collision avoidance and task optimization. They are extensively described in this section as they are critically important for the overall system.

3.2.1 Resolved Motion Rate Control

The kinematic output of a generic robotic manipulator is usually represented by a *manipulation variable*, $\boldsymbol{r} \in \Re^m$. A manipulation variable may be, for example, the position and orientation of the end-effector, the measure of manipulability, and any other functions of the joint variable, $\boldsymbol{q}$:

$$\boldsymbol{r} = f\,(\boldsymbol{q})\,. \tag{3.5}$$

Considering small variations, the relationship between $\delta \boldsymbol{r}$ and $\delta \boldsymbol{q}$ is given by:

$$\delta \boldsymbol{r} = \frac{\partial f}{\partial \boldsymbol{q}} \delta \boldsymbol{q} = \boldsymbol{J}(\boldsymbol{q})\, \delta \boldsymbol{q} \tag{3.6}$$

where $\boldsymbol{J}(\boldsymbol{q}) \in \Re^{m \times n}$ is the Jacobian matrix of the manipulation variable, $\boldsymbol{r}$. In resolved motion rate control [3], we compute $\delta \boldsymbol{q}$ for a given $\delta \boldsymbol{r}$ and $\boldsymbol{q}$ by solving the linear system, Eq. (4.28). In the general case, this is done by using the least-square solution, which is $\delta \boldsymbol{q}$ that minimizes the error norm:

$$min\|\delta \boldsymbol{r} - \boldsymbol{J}(\boldsymbol{q})\, \delta \boldsymbol{q}\| \tag{3.7}$$

The solution of Eq. (3.7) is given in the general form by [4]:

$$\delta \boldsymbol{q} = \boldsymbol{J}^{+}(\boldsymbol{q})\, \delta \boldsymbol{r} + \left(\boldsymbol{I}_n - \boldsymbol{J}^{+}(\boldsymbol{q})\, \boldsymbol{J}(\boldsymbol{q})\right) \boldsymbol{y} \tag{3.8}$$

where $\boldsymbol{J}^{+}(\boldsymbol{q}) \in \Re^{n \times m}$ is the Moore-Penrose pseudo-inverse of $\boldsymbol{J}(\boldsymbol{q})$, $\boldsymbol{y} \in \Re^{n}$ is an arbitrary vector and $\boldsymbol{I}_n \in \Re^{n \times n}$ indicates an identity matrix.

The least-square solution is not necessarily unique. This happens when the manipulator possesses more degrees of freedom than the dimension of m of the manipulation variable $\boldsymbol{r}$. In this case, the manipulator may be considered kinematically redundant with respect to the particular task defined by Eq. (3.5), and every solution is obtained with different values of $\boldsymbol{y}$ in Eq. (3.8). Among all its solutions, the one that also minimizes the norm $\|\delta \boldsymbol{q}\|$ is obtained with $\boldsymbol{y} = \boldsymbol{0}$:

$$\delta \boldsymbol{q} = \boldsymbol{J}^{+}(\boldsymbol{q})\, \delta \boldsymbol{r} \tag{3.9}$$

When there is no interest in using the redundancy for attaining secondary goals, this is usually the solution preferred because it minimizes the joint velocity.

If the kinematic equation describes a non-redundant case ($m = n$), $\boldsymbol{J}^{+}(\boldsymbol{q})$ is equal to $\boldsymbol{J}^{-1}(\boldsymbol{q})$ and $\left(\boldsymbol{I} - \boldsymbol{J}^{+}(\boldsymbol{q})\, \boldsymbol{J}(\boldsymbol{q})\right)$ becomes $\boldsymbol{0}$.

Note that, in general, the first and second terms of Eq. (3.8) are orthogonal:

$$\begin{aligned} \left(\boldsymbol{J}^{+} \delta \boldsymbol{r}\right)^{T} \left(\boldsymbol{I}_n - \boldsymbol{J}^{+}\boldsymbol{J}\right) \boldsymbol{y} &= \delta \boldsymbol{r}^{T} \left(\boldsymbol{J}^{+}\right)^{T} \left(\boldsymbol{I}_n - \boldsymbol{J}^{+}\boldsymbol{J}\right) \boldsymbol{y} \\ &= \delta \boldsymbol{r}^{T} \left[\left(\boldsymbol{I}_n - \boldsymbol{J}^{+}\boldsymbol{J}\right) \boldsymbol{J}^{+}\right]^{T} \boldsymbol{y} \\ &= \boldsymbol{0} \end{aligned} \tag{3.10}$$

where the symmetry of $\boldsymbol{I}_n - \boldsymbol{J}^{+}\boldsymbol{J}$ and the *Penrose* condition $\boldsymbol{J}^{+}\boldsymbol{J}\boldsymbol{J}^{+} = \boldsymbol{J}^{+}$ have been used [5].

3.2.1.1 Kinematic Singularities

A *kinematic singularity* is defined as the joint configuration vector value $\boldsymbol{q}^*$ such that $\boldsymbol{J}(\boldsymbol{q}^*)$ does not have full rank. Its pseudo-inverse, $\boldsymbol{J}^+(\boldsymbol{q}^*)$, is not defined at such a configuration. Physically, in the neighborhood of a singular configuration, even a small change in $\delta\boldsymbol{r}$ requires an enormous change in $\delta\boldsymbol{q}$, which is not practically feasible in real manipulators and also dangerous for the structure.

The *damped least-squares method* [6] is a classical and simple way to overcome this drawback. It consists in adding a regularization term acting in the neighborhood of the singularities:

$$\boldsymbol{J}^+_{DLS} = \boldsymbol{J}^T\left(\boldsymbol{J}\boldsymbol{J}^T + \lambda\boldsymbol{I}\right)^{-1}. \tag{3.11}$$

Consequently, the damped least square solution of Eq. (4.28) is:

$$\delta\boldsymbol{q}_{DLS} = \boldsymbol{J}^+_{DLS}\delta\boldsymbol{r} + \left[\boldsymbol{I}_n - \boldsymbol{J}^+_{DLS}(\boldsymbol{q})\boldsymbol{J}(\boldsymbol{q})\right]\boldsymbol{y} \tag{3.12}$$

However, the main disadvantage for this approach is a loss of performance and an increased tracking error [7]. The choice of the damping terms must balance the required performance with the error allowed. To overcome those defects, Nakamura [8] introduced a variable damping factor (*singularity-robust inverse*), with the regularization terms acting only in proximity of the singular points. Chiaverini also proposed a modified inverse, adding only the damping parameter to the lowest singular values [7]. These approaches are considerably better than the damped least-squares method. However all the above solutions present the common problem of tuning parameters, and more important in autonomous manipulation, there is a certain level of unpredictability in the movements when the manipulator falls within a singular configuration.

3.2.2 Task-Priority-Based Decomposition

In autonomous robotic systems, the subtask decomposition between position and orientation is advantageous, because it will enlarge the reachable workspace of the first-priority manipulation variable (usually position) by allowing incompleteness for the second priority subtask. The concept of task priority was introduced by Nakamura [4] into the inverse kinematics of manipulators. Let the manipulation variable $\boldsymbol{r}_1 \in \Re^{m_1}$ be our first priority task:

$$\boldsymbol{r}_1 = f_1(\boldsymbol{q}), \tag{3.13}$$

where $\boldsymbol{q} \in \Re^n$ is the robot configuration vector and $\boldsymbol{r}_1$ can be, for example, the position of the end-effector. The differential relationship of (3.13) is:

$$\delta\boldsymbol{r}_1 = \boldsymbol{J}_1(\boldsymbol{q})\,\delta\boldsymbol{q} \tag{3.14}$$

where $\boldsymbol{J}_1(\boldsymbol{q}) \in \Re^{m_1 \times n}$ is the Jacobian matrix of the first manipulation variable, $\boldsymbol{r}_1$. Likewise, if we have additional degrees of freedom, let the manipulation variable $\boldsymbol{r}_2 \in \Re^{m_2}$ be our secondary task:

$$\boldsymbol{r}_2 = f_2(\boldsymbol{q}) \tag{3.15}$$

$$\delta\boldsymbol{r}_2 = \boldsymbol{J}_2(\boldsymbol{q})\,\delta\boldsymbol{q} \tag{3.16}$$

where $\boldsymbol{J}_2(\boldsymbol{q}) \in \Re^{m_2 \times n}$ is the Jacobian matrix of the secondary task, $\boldsymbol{r}_2$. Equation (3.14) has an infinite variety of solutions for $\delta\boldsymbol{q}$, whose general solution is obtained using the pseudoinverse solution of the Jacobian matrix:

$$\delta\boldsymbol{q} = \boldsymbol{J}_1{}^{+}(\boldsymbol{q})\,\delta\boldsymbol{r}_1 + \left[\boldsymbol{I}_n - \boldsymbol{J}_1{}^{+}(\boldsymbol{q})\,\boldsymbol{J}_1(\boldsymbol{q})\right]\boldsymbol{y} \tag{3.17}$$

where $\boldsymbol{J}_1{}^{+}(\boldsymbol{q}) \in \Re^{n \times m_1}$ is the pseudoinverse of $\boldsymbol{J}_1(\boldsymbol{q})$, $\boldsymbol{y} \in \Re^{n}$ is an arbitrary vector and $\boldsymbol{I}_n \in \Re^{n \times n}$ indicates an identity matrix. Substituting Eq. (3.17) into Eq. (3.16), we obtain:

$$\boldsymbol{J}_2\left(\boldsymbol{I}_n - \boldsymbol{J}_1{}^{+}\boldsymbol{J}_1\right)\boldsymbol{y} = \delta\boldsymbol{r}_2 - \boldsymbol{J}_2\boldsymbol{J}_1{}^{+}\delta\boldsymbol{r}_1. \tag{3.18}$$

If the exact solution of $\boldsymbol{y}$ exists, Eq. (3.18) implies that the second manipulation variable can be realized. Generally, the exact solution does not exist, however, we can obtain $\boldsymbol{y}$ that minimizes $\left\|\delta\boldsymbol{r}_2 - \boldsymbol{J}_2\boldsymbol{J}_1{}^{+}\delta\boldsymbol{r}_1\right\|$ in the least square sense by using again the pseudo-inverse:

$$\boldsymbol{y} = \hat{\boldsymbol{J}}_2^{+}\left(\delta\boldsymbol{r}_2 - \boldsymbol{J}_2\boldsymbol{J}_1{}^{+}\delta\boldsymbol{r}_1\right) + \left(\boldsymbol{I}_n - \hat{\boldsymbol{J}}_2^{+}\hat{\boldsymbol{J}}_2\right)\boldsymbol{z}, \tag{3.19}$$

$$\hat{\boldsymbol{J}}_2 = \boldsymbol{J}_2\left(\boldsymbol{I}_n - \boldsymbol{J}_1{}^{+}\boldsymbol{J}_1\right). \tag{3.20}$$

where $\boldsymbol{z} \in \Re^{n}$ is an arbitrary vector.

Finally, substituting Eq. (3.19) into Eq. (3.17), we obtain:

$$\delta\boldsymbol{q} = \boldsymbol{J}_1{}^{+}\delta\boldsymbol{r}_1 + \hat{\boldsymbol{J}}_2^{+}\left(\delta\boldsymbol{r}_2 - \boldsymbol{J}_2\boldsymbol{J}_1{}^{+}\delta\boldsymbol{r}_1\right) + \left(\boldsymbol{I}_n - \boldsymbol{J}_1{}^{+}\boldsymbol{J}_1\right)\left(\boldsymbol{I}_n - \hat{\boldsymbol{J}}_2^{+}\hat{\boldsymbol{J}}_2\right)\boldsymbol{z}. \tag{3.21}$$

If we still have remaining redundancy, let now introduce a third manipulation variable $\boldsymbol{r}_3 \in \Re^{m_3}$:

$$\boldsymbol{r}_3 = f_3(\boldsymbol{q}), \tag{3.22}$$

$$\delta\boldsymbol{r}_3 = \boldsymbol{J}_3(\boldsymbol{q})\,\delta\boldsymbol{q}. \tag{3.23}$$

Using again the above procedure, we obtain:

$$\delta\boldsymbol{q} = \boldsymbol{J}_1{}^{+}\delta\boldsymbol{r}_1 + \hat{\boldsymbol{J}}_2^{+}\left(\delta\boldsymbol{r}_2 - \boldsymbol{J}_2\delta\boldsymbol{q}_1\right) + \hat{\boldsymbol{J}}_3^{+}\left[\delta\boldsymbol{r}_3 - \boldsymbol{J}_3\left(\delta\boldsymbol{q}_1 + \delta\boldsymbol{q}_2\right)\right] \tag{3.24}$$

where:

$$\hat{\boldsymbol{J}}_2 = \boldsymbol{J}_2 \left(\boldsymbol{I}_n - \boldsymbol{J}_1^+ \boldsymbol{J}_1 \right), \tag{3.25}$$

$$\hat{\boldsymbol{J}}_3 = \boldsymbol{J}_3 \left(\boldsymbol{I}_n - \boldsymbol{J}_1^+ \boldsymbol{J}_1 - \hat{\boldsymbol{J}}_2^+ \hat{\boldsymbol{J}}_2 \right). \tag{3.26}$$

Equation (3.24) suggests the recursive idea:

$$\begin{cases} \delta \boldsymbol{q}_i = \delta \boldsymbol{q}_{i-1} + \hat{\boldsymbol{J}}_i^+ \left(\delta \boldsymbol{r}_i - \boldsymbol{J}_i \delta \boldsymbol{q}_{i-1} \right) \\ \hat{\boldsymbol{J}}_i = \boldsymbol{J}_i \boldsymbol{J}_i{}^n \\ \boldsymbol{J}_i{}^n = \boldsymbol{J}_{i-1}{}^n - \hat{\boldsymbol{J}}_{i-1}^+ \hat{\boldsymbol{J}}_{i-1} \end{cases}, \quad \begin{cases} \delta \boldsymbol{q}_0 = \boldsymbol{0} \\ \boldsymbol{J}_0 = \boldsymbol{0} \\ \boldsymbol{J}_0{}^n = \boldsymbol{I}_n \end{cases} \tag{3.27}$$

also addressed in [9, 10].

3.2.2.1 Algorithmic Singularities

In the task decomposition approach, the occurrence of algorithmic singularities arises from conflicts between the different tasks, when the correspondent non-prioritized task is not feasible. Mathematically, in case of two manipulation variables, algorithmic singularities are configurations at which the matrix $\hat{\boldsymbol{J}}_2$ in Eq. (3.25) loses rank with $\boldsymbol{J}_1$ and $\boldsymbol{J}_2$ of full rank.

Only a few attempts have been made for the avoidance of algorithmic singularities as shown in [7, 11, 12]. Chiaverini's optimization method [7] is free from the algorithmic singularity, at the expenses of an increased secondary task error. To increase task performance, the Bordered Grammian method was proposed in [12]. However, this method introduces another kind of singularity and its performance is somewhat limited.

The *Task Reconstruction* for singularity avoidance [13–16] presented in the following section, provides a valid solution to avoid both kinematic and algorithmic singularities, and by our experiments was proved to be the most effective approach in case of autonomous manipulation.

3.2.3 Measure of Manipulability

The main requirement in a singularity avoidance approach is the localization of a singular configuration within the joint space. Yoshikawa [17] proposed a continuous measure that evaluates the kinematic quality of robot mechanism:

$$\mu \left(\boldsymbol{J} \right) = \sqrt{\det \left(\boldsymbol{J} \boldsymbol{J}^T \right)}. \tag{3.28}$$

$\mu \left(\boldsymbol{J} \right)$ takes a continuous non-negative scalar value and becomes equal to zero only when the Jacobian matrix is not full rank.

Let's consider the singular value decomposition of the Jacobian matrix $\boldsymbol{J}$:

$$\boldsymbol{J}(\boldsymbol{q}) = \boldsymbol{U}\boldsymbol{\Sigma}\boldsymbol{V}^T, \tag{3.29}$$

where $\boldsymbol{U}$ and $\boldsymbol{V}$ are orthogonal matrixes and $\boldsymbol{\Sigma}$ is a diagonal matrix whose diagonal elements are the ordered singular values of $\boldsymbol{J}$:

$$\boldsymbol{\Sigma} = \begin{cases} \left[\operatorname{diag}(\sigma_1, \ldots, \sigma_m) \mid \boldsymbol{0},\right], & (m \leq n), \\ \begin{bmatrix} \operatorname{diag}(\sigma_1, \ldots, \sigma_n) \\ \boldsymbol{0} \end{bmatrix}, & (m > n). \end{cases} \tag{3.30}$$

Substituting Eq. (3.29) into Eq. (3.28) results in:

$$\begin{aligned} \mu &= \sqrt{\det\left(\boldsymbol{U}\boldsymbol{\Sigma}\boldsymbol{V}^T\boldsymbol{V}\boldsymbol{\Sigma}^T\boldsymbol{U}^T\right)} \\ &= \sqrt{\det\left(\boldsymbol{\Sigma}\boldsymbol{\Sigma}^T\right)} = \prod_{i=1}^{n} |\sigma_i| . \end{aligned} \tag{3.31}$$

Equation (3.31) shows that $\mu(\boldsymbol{J})$ is exactly the product of the singular values of $\boldsymbol{J}$ and can be regarded as a distance from singularity.

3.2.3.1 Derivative of Measure of Manipulability

One advantage of this choice as a distance from singularity is that we can find its derivative with respect to the joint configuration vector $\boldsymbol{q}$, which is needed for the controller. Indicating with $j(i, j)$ the element (i, j) of the Jacobian matrix $\boldsymbol{J}$, the derivative of $\mu(\boldsymbol{J})$ with respect to the i-th component of $\boldsymbol{q}$ is given by:

$$\frac{\partial \mu(\boldsymbol{J})}{\partial q_k} = \sum_{i,j} \frac{\partial \mu(\boldsymbol{J})}{\partial j(i,j)} \frac{\partial j(i,j)}{\partial q_k} = \sum_{i,j} \left[\frac{\partial \mu(\boldsymbol{J})}{\partial \boldsymbol{J}}\right]_{(i,j)} \frac{\partial j(i,j)}{\partial q_k} \tag{3.32}$$

where $\frac{\partial f}{\partial \boldsymbol{X}}$ is a matrix whose (i, j) element is $\frac{\partial f}{\partial x_{i,j}}$. Thus:

$$\begin{aligned} \frac{\partial \mu(\boldsymbol{J})}{\partial q_k} &= \sum_{i,j} \left[\frac{1}{2\sqrt{\det\left(\boldsymbol{J}\boldsymbol{J}^T\right)}} \frac{\partial \det\left(\boldsymbol{J}\boldsymbol{J}^T\right)}{\partial \boldsymbol{J}^T}\right]^T_{(i,j)} \frac{\partial j(i,j)}{\partial q_k} \\ &= \sum_{i,j} \left[\frac{1}{2\sqrt{\det\left(\boldsymbol{J}\boldsymbol{J}^T\right)}} 2\det\left(\boldsymbol{J}\boldsymbol{J}^T\right)\boldsymbol{J}^T\left(\boldsymbol{J}\boldsymbol{J}^T\right)^{-1}\right]^T_{(i,j)} \frac{\partial j(i,j)}{\partial q_k} \end{aligned} \tag{3.33}$$

where the relation:

$$\frac{d}{d\boldsymbol{X}}\left(\det\left(\boldsymbol{X}^T\boldsymbol{X}\right)\right) = 2\det\left(\boldsymbol{X}^T\boldsymbol{X}\right)\cdot\boldsymbol{X}\left(\boldsymbol{X}^T\boldsymbol{X}\right)^{-1} \tag{3.34}$$

has been used. Assuming that $\boldsymbol{J}$ is of full rank we have:

$$\frac{\partial mom\,(\boldsymbol{J})}{\partial q_k} = \sum_{i,j}\left[\sqrt{\det\left(\boldsymbol{J}\boldsymbol{J}^T\right)}\boldsymbol{J}^+\right]^T_{(i,j)}\frac{\partial j\,(i,j)}{\partial q_k} \tag{3.35}$$

Finally:

$$\frac{\partial\mu\,(\boldsymbol{J})}{\partial q_k} = \mu\,(\boldsymbol{J})\sum_{i,j}\left[\boldsymbol{J}^{+T}\right]_{(i,j)}\left[\frac{\partial\boldsymbol{J}}{\partial q_k}\right]_{(i,j)} \tag{3.36}$$

$$\frac{\partial\mu\,(\boldsymbol{J})}{\partial q_k} = \mu\,(\boldsymbol{J})\cdot\text{trace}\left\{\frac{\partial\boldsymbol{J}}{\partial q_k}\boldsymbol{J}^+\right\}. \tag{3.37}$$

Equation (3.37) shows that we can express the derivative of measure of manipulability with respect to some already known quantities, as $\mu\,(\boldsymbol{J})$ itself and the pseudo-inverse of Jacobian matrix, $\boldsymbol{J}^+$. The derivative of each element of $\boldsymbol{J}$ with respect to q_k may be easily computed symbolically. The cost of its numerical computation is lower than the one for the Jacobian itself, and is the only one added for computing $\frac{\partial\mu(\boldsymbol{J})}{\partial q_k}$. The above formulation has been also discussed in [18].

3.3 Task Reconstruction for Singularity Avoidance

For a given manipulation variable, usually a singularity-free motion path may be achieved with an off-line path planning. However, this approach requires preliminary knowledge of all the singular configurations of the manipulator, which is not always possible for large DOF systems.

The proposed method, based on a real-time evaluation of the measure of manipulability, allows moving along a singularity-free path for a generic manipulator whose singular configurations are not known in advance. The basic idea is to *reconstruct* the main task *when* approaching to a singular configuration, in order to move along a path where the distance from the singularity point is maintained constant (or at least not decreasing). This approach is particularly suitable in autonomous manipulation because it releases the task generator (the higher layer in the main control system) from keeping into account the singularity problem.

3.3.1 Task Reconstruction: Single Manipulation Variable

The concept of task reconstruction has been introduced in [14, 15, 19, 20] and successively generalized in [13].

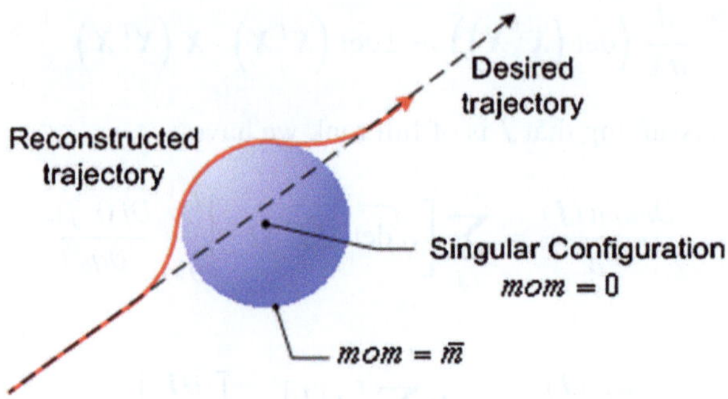

Fig. 3.4 Conceptual diagram of the task reconstruction method

Figure 3.4 shows the main idea, which consists basically in modifying the main task in order to move along the path where the distance from singularity (*mom*) is maintained over a predefined value.

Let's consider again the solution of inverse kinematics in case of one manipulation variable (3.8):

$$\delta \boldsymbol{q} = \boldsymbol{J}^{+}(\boldsymbol{q})\, \delta \boldsymbol{r} + \left(\boldsymbol{I}_n - \boldsymbol{J}^{+}(\boldsymbol{q})\, \boldsymbol{J}(\boldsymbol{q})\right) \boldsymbol{y} \tag{3.38}$$

In the above equation, let's consider $\boldsymbol{y} = \boldsymbol{0}$ (absence of null motion):

$$\delta \boldsymbol{q} = \boldsymbol{J}^{+}(\boldsymbol{q})\, \delta \boldsymbol{r} \tag{3.39}$$

In the following treatise we can assume that $\boldsymbol{J}(\boldsymbol{q})$ is always of full rank. As a matter of fact, this is the goal of our singularity avoidance approach.

Let's introduce now a function $m(\boldsymbol{q})$ of the configuration vector $\boldsymbol{q}$ that represents the measure of the distance of the system from a singular configuration. In our approach, we chose the measure of manipulability introduced in Eq. (3.28):

$$m(\boldsymbol{q}) = mom(\boldsymbol{q}) \tag{3.40}$$

The small variation of the measure $m(\boldsymbol{q})$ is given by:

$$\delta m(\boldsymbol{q}) = \frac{\partial m(\boldsymbol{q})}{\partial \boldsymbol{q}} \delta \boldsymbol{q} = \frac{\partial m(\boldsymbol{q})}{\partial \boldsymbol{q}} \boldsymbol{J}^{+} \delta \boldsymbol{r}. \tag{3.41}$$

In order to have $\delta m(\boldsymbol{q}) = 0$, Eq. 3.41 implies that the given task must be orthogonal to the vector:

$$\frac{\partial m(\boldsymbol{q})}{\partial \boldsymbol{q}} \boldsymbol{J}^{+} \tag{3.42}$$

or, equivalently, that $\delta \boldsymbol{r}$ must lie on the surface defined by:

$$\left\{ \boldsymbol{x} \in \Re^m \; : \; \left(\frac{\partial m\left(\boldsymbol{q}\right)}{\partial \boldsymbol{q}} \boldsymbol{J}^+ \right) \cdot \boldsymbol{x} = 0 \right\} \tag{3.43}$$

Let $\boldsymbol{n}_m$ be the unitary vector orthogonal to the surface defined by Eq. 3.43:

$$\boldsymbol{n}_m = \frac{\left(\frac{\partial m(\boldsymbol{q})}{\partial \boldsymbol{q}} \boldsymbol{J}^+ \right)^T}{\left\| \frac{\partial m(\boldsymbol{q})}{\partial \boldsymbol{q}} \boldsymbol{J}^+ \right\|} \tag{3.44}$$

Consequently, the projection of the given task on the surface is:

$$\begin{aligned} \delta \boldsymbol{r}_p &= \delta \boldsymbol{r} - \left(\delta \boldsymbol{r} \cdot \boldsymbol{n}_m \right) \boldsymbol{n}_m \\ &= \delta \boldsymbol{r} - \left(\boldsymbol{n}_m \boldsymbol{n}_m{}^T \right) \delta \boldsymbol{r} \\ &= \left(\boldsymbol{I}_m - \boldsymbol{n}_m \boldsymbol{n}_m{}^T \right) \delta \boldsymbol{r}. \end{aligned} \tag{3.45}$$

This projection eliminates the components of the task that may drive the system toward the singular configuration.[1] This action may occur only when the distance from singularity is less than or equal to a predefined value. At that aim, it is necessary to introduce a weight into Eq. (3.45) as follows:

$$\delta \boldsymbol{r}_p = \left(\boldsymbol{I}_m - \boldsymbol{n}_m \boldsymbol{n}_m{}^T k_m \left(m, \overline{m}, \sigma \right) \right) \delta \boldsymbol{r}, \tag{3.46}$$

where $k_m\left(m, \overline{m}, \sigma\right)$ is a positive and well-shaped function, which is equal to 1 for values of $m\left(\boldsymbol{q}\right)$ smaller than the boundary of the singular region, $\overline{m}$, and equal to zero for values of $m\left(\boldsymbol{q}\right)$ greater than the boundary $\overline{m}$ plus a transitional zone σ (with continuous first derivative). The shape function is defined as:

$$k_m\left(m, \overline{m}, \sigma\right) = \begin{cases} 1. & m \leq \overline{m} \\ 2\left(\frac{m-\overline{m}}{\sigma}\right)^3 - 3\left(\frac{m-\overline{m}}{\sigma}\right)^2 + 1, & \overline{m} < m \leq \overline{m} + \sigma \\ 0, & \overline{m} + \sigma < m \end{cases} \tag{3.47}$$

Figure 3.5 shows an example of Eq. (3.47) for $\overline{m} = \sigma = 0.02$. The continuity of the first derivative allows to progressively lay down the task solution, $\delta \boldsymbol{r}$, on the surface where $m\left(\boldsymbol{q}\right)$ is constant, without introducing instabilities to the controller when closing the loop.

In addition, we must ensure to leave from the surface by acting the task correction in Eq. (3.46) only when the scalar product $\delta \boldsymbol{r} \cdot \boldsymbol{n}_m$ is negative, that is when the singularity measure $m\left(\boldsymbol{q}\right)$ is decreasing:

[1] In case that $\left\| \frac{\partial m(\boldsymbol{q})}{\partial \boldsymbol{q}} \boldsymbol{J}^+ \right\| = 0$, the normal (3.44) is not defined. However this means that, locally, $\delta m\left(\boldsymbol{q}\right) = 0$ in every direction, thus the value of $\boldsymbol{n}_m$ is not important.

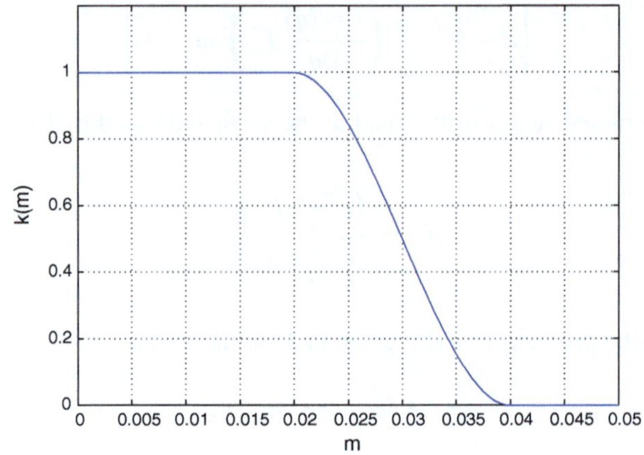

Fig. 3.5 Shape function

$$\begin{aligned}\delta\boldsymbol{r}_p &= \left(\boldsymbol{I}_m - \frac{1-\text{sign}\left(\delta\boldsymbol{r}\cdot\boldsymbol{n}_m\right)}{2}\left(\boldsymbol{n}_m\boldsymbol{n}_m{}^T\right)k_m\left(m,\overline{m},\sigma\right)\right)\delta\boldsymbol{r},\\ &= \left(\boldsymbol{I}_m - k_1\left(\boldsymbol{n}_m\boldsymbol{n}_m{}^T\right)\right)\delta\boldsymbol{r},\end{aligned} \tag{3.48}$$

where $k_1 = \frac{1}{2}\left(1 - sign\left(\delta\boldsymbol{r}\cdot\boldsymbol{n}_m\right)\right)k_m\left(m,\overline{m},\sigma\right)$. When $m\left(\boldsymbol{q}\right)$ is already smaller than the value on the surface, Eq. (3.48) does not guarantee to escape from the enclosed volume. Furthermore, numerical errors may introduce a small drift term driving the task below the surface. To avoid the above drawbacks, one additional term is introduced into Eq. (3.48):

$$\delta\boldsymbol{r}_p = \left(\boldsymbol{I}_m - k_1\boldsymbol{n}_m\boldsymbol{n}_m{}^T\right)\delta\boldsymbol{r} + k_2\boldsymbol{n}_m, \tag{3.49}$$

$$k_2(m) = K_r k_m\left(m, \frac{\overline{m}}{2}, \frac{\overline{m}}{2}\right) \tag{3.50}$$

where K_r is a scalar gain. With the above choice of $\overline{m}$ and σ, k_2 is different from zero only for $m\left(\boldsymbol{q}\right) < \overline{m}$. The effect is a recalling action toward the surface, starting when $\delta\boldsymbol{r}_p$ has no more components along the gradient. Figure 3.6 shows the geometric interpretation of the TR method.

In our implementation chose $\sigma = \overline{m}$. Thus the only tunable parameters are K_r and the minimum allowed distance from singularity $\overline{m}$. The last parameter, $\overline{m}$, is responsible for the performance of the system when working close to a singularity. A small value is preferable; however getting too close to the singularity may be dangerous for the presence of numerical errors and noise.

The above reconstructed task $\delta\boldsymbol{r}_p$ is used in Eq. (3.39) in place of the original task $\delta\boldsymbol{r}$, guaranteeing a total singularity-free motion path within the arm workspace. Thus, with the above choice, we have:

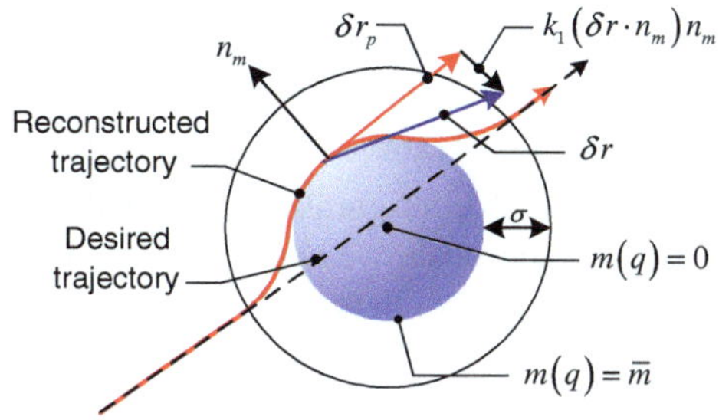

Fig. 3.6 Geometric interpretation of the task reconstruction concept

$$\delta \boldsymbol{q} = \boldsymbol{J}^{+}(\boldsymbol{q})\, TR(\boldsymbol{J}, m(\boldsymbol{q}), \delta \boldsymbol{r}) \tag{3.51}$$

where:

$$\begin{aligned}
TR(\boldsymbol{J}, m(\boldsymbol{q}), \delta \boldsymbol{r}) &= \left(\boldsymbol{I}_m - k_1 \boldsymbol{n}_m \boldsymbol{n}_m{}^T\right) \delta \boldsymbol{r} + k_2 \boldsymbol{n}_m \\
\boldsymbol{n}_m &= \frac{\left(\frac{\partial m(\boldsymbol{q})}{\partial \boldsymbol{q}} \boldsymbol{J}^{+}\right)^T}{\left\| \frac{\partial m(\boldsymbol{q})}{\partial \boldsymbol{q}} \boldsymbol{J}^{+} \right\|} \\
k_1 &= \frac{1 - \operatorname{sign}(\delta \boldsymbol{r} \cdot \boldsymbol{n}_m)}{2} k_m(m, \overline{m}, \overline{m}) \\
k_2 &= K_r k_m \left(m, \frac{\overline{m}}{2}, \frac{\overline{m}}{2}\right)
\end{aligned} \tag{3.52}$$

Equation (3.52) represents the general form of our task reconstruction process, indicated for simplicity as a function $TR(\boldsymbol{J}, m(\boldsymbol{q}), \delta \boldsymbol{r})$ of the Jacobian $\boldsymbol{J}$, the measure $m(\boldsymbol{q})$ and the variation of the manipulation variable $\delta \boldsymbol{r}$.

3.3.2 Task Reconstruction: Case of Two Tasks with Order of Priority

Let's consider the case of two subtasks with absence of null motion. The inverse kinematics taking into account of the order of priority is given by Eq. (3.21) for $\boldsymbol{z} = \boldsymbol{0}$:

$$\delta \boldsymbol{q} = \boldsymbol{J}_1{}^{+} \delta \boldsymbol{r}_1 + \hat{\boldsymbol{J}}_2^{+} \delta \hat{\boldsymbol{r}}_2 \tag{3.53}$$

with:

$$\delta \hat{\boldsymbol{r}}_2 = \delta \boldsymbol{r}_2 - \boldsymbol{J}_2 \boldsymbol{J}_1^{+} \delta \boldsymbol{r}_1 \tag{3.54}$$

$$\hat{\boldsymbol{J}}_2 = \boldsymbol{J}_2 \left(\boldsymbol{I}_n - \boldsymbol{J}_1{}^{+} \boldsymbol{J}_1\right). \tag{3.55}$$

In this case, the system may also be affected by algorithmic singularities. As seen before, they are configurations at which the matrix $\hat{J}_2$ in Eq. (3.55) loses rank with J_1 and J_2 of full rank. A loss of rank of $\hat{J}_2$ implies, from Eq. (3.31), that $\mu\left(\hat{J}_2\right) = 0$. Avoiding such configuration will result in avoiding also algorithmic singularities.

In our approach, we apply the task reconstruction process (3.52) to the modified secondary task $\delta\hat{r}_2$ in order to prevent $\hat{J}_2$ to lose rank. This is done after applying the TR process (3.52) to the first task in order to prevent kinematic singularities. Thus we simply have:

$$\delta q = J_1{}^{+}\delta r_{1p} + \hat{J}_2^{+}\delta\hat{r}_{2p} \tag{3.56}$$

where:

$$r_{1p} = TR\left(J_1, m_1(q), \delta r_1\right) \tag{3.57}$$

$$r_{2p} = TR\left(\hat{J}_2, m_2(q), \delta r_2 - J_2 J_1^{+}\delta r_{1p}\right) \tag{3.58}$$

The measures of the distance from singularity $m_1(q)$ and $m_2(q)$ are given by Eq. (3.28):

$$m_1(q) = \mu\left(J_1\right) \tag{3.59}$$

$$m_2(q) = \mu\left(\hat{J}_2\right) \tag{3.60}$$

In the above formulation, while $\frac{\partial m(q)}{\partial q}$ can be easily computed in case of the first task using the Eq. (3.37), in case of the secondary task the computation of $\frac{\partial\mu\left(\hat{J}_2\right)}{\partial q}$ is not immediate.

At this aim let's consider the product of the measure of manipulability of J_1 and $\hat{J}_2$. From Eq. (3.31) we have:

$$\mu\left(J_1\right)\cdot\mu\left(\hat{J}_2\right) = \sqrt{\det\left(J_1 J_1{}^{T}\right)\cdot\det\left(\hat{J}_2\hat{J}_2^{T}\right)}. \tag{3.61}$$

From Eq. (3.20) we have:

$$\begin{aligned}\det\left(\hat{J}_2\hat{J}_2^{T}\right) &= \det\left(J_2\left(I_n - J_1{}^{+}J_1\right)\left(I_n - J_1{}^{+}J_1\right)^{T} J_2{}^{T}\right)\\ &= \det\left(J_2 J_2{}^{T} - J_2 J_1{}^{+} J_1 J_2{}^{T}\right)\\ &= \det\left(J_2 J_2{}^{T} - J_2 J_1{}^{T}\left(J_1 J_1{}^{T}\right)^{-1} J_1 J_2{}^{T}\right)\end{aligned} \tag{3.62}$$

using the idempotency of $\left(I_n - J_1{}^{+}J_1\right)$. Recalling the determinant formula for the 2-by-2 block matrix:

$$\det\left(\begin{bmatrix} \boldsymbol{A} & \boldsymbol{B} \\ \boldsymbol{C} & \boldsymbol{D} \end{bmatrix}\right) = \det(\boldsymbol{A}) \cdot \det\left(\boldsymbol{D} - \boldsymbol{C}\boldsymbol{A}^{-1}\boldsymbol{B}\right) \tag{3.63}$$

we obtain [21]:

$$\begin{aligned} \mu(\boldsymbol{J}_1) \cdot \mu\left(\hat{\boldsymbol{J}}_2\right) &= \sqrt{\det\left(\begin{bmatrix} \boldsymbol{J}_1\boldsymbol{J}_1{}^T & \boldsymbol{J}_1\boldsymbol{J}_2{}^T \\ \boldsymbol{J}_2\boldsymbol{J}_1{}^T & \boldsymbol{J}_2\boldsymbol{J}_2{}^T \end{bmatrix}\right)} \\ &= \mu\left(\begin{bmatrix} \boldsymbol{J}_1 \\ \boldsymbol{J}_2 \end{bmatrix}\right) \end{aligned} \tag{3.64}$$

or, equivalently:

$$\sqrt{\det\left(\boldsymbol{J}_1\boldsymbol{J}_1{}^T\right)} \cdot \sqrt{\det\left(\hat{\boldsymbol{J}}_2\hat{\boldsymbol{J}}_2^T\right)} = \sqrt{\det\left(\boldsymbol{J}\boldsymbol{J}^T\right)} \tag{3.65}$$

where:

$$\boldsymbol{J} = \begin{bmatrix} \boldsymbol{J}_1 \\ \boldsymbol{J}_2 \end{bmatrix} \tag{3.66}$$

is the matrix obtained by stacking $\boldsymbol{J}_1$ and $\boldsymbol{J}_2$. Equation (3.65) means that algorithmic singularities in the inverse kinematics taking account of the priority of the subtasks will not occur when the Jacobian of the corresponding task-space augmentation approach:

$$\delta\boldsymbol{r} = \begin{bmatrix} \delta\boldsymbol{r}_1 \\ \delta\boldsymbol{r}_2 \end{bmatrix} = \begin{bmatrix} \boldsymbol{J}_1(\boldsymbol{q}) \\ \boldsymbol{J}_2(\boldsymbol{q}) \end{bmatrix} \delta\boldsymbol{q} \tag{3.67}$$

is not singular. More precisely [21], a singular configuration in the augmented Jacobian (3.67) occurs when $\boldsymbol{J}_1$ or $\boldsymbol{J}_2$ are rank deficient or the primary and secondary task are incompatible.

In order to compute the derivative of $m_2(\boldsymbol{q})$ with respect the configuration vector $\boldsymbol{q}$, we can use Eq. (3.64):

$$\mu(\boldsymbol{J}) = \mu(\boldsymbol{J}_1) \cdot \mu\left(\hat{\boldsymbol{J}}_2\right) \tag{3.68}$$

Deriving both sides of Eq. (3.68) we obtain:

$$\frac{d\mu(\boldsymbol{J})}{d\boldsymbol{q}} = \frac{d\mu(\boldsymbol{J}_1)}{d\boldsymbol{q}}\mu\left(\hat{\boldsymbol{J}}_2\right) + \frac{d\mu\left(\hat{\boldsymbol{J}}_2\right)}{d\boldsymbol{q}}\mu(\boldsymbol{J}_1) \tag{3.69}$$

that can be solved with respect to $\frac{d\mu\left(\hat{\boldsymbol{J}}_2\right)}{d\boldsymbol{q}}$:

$$\frac{d\boldsymbol{m}_2(\boldsymbol{q})}{d\boldsymbol{q}} = \frac{d\mu\left(\hat{\boldsymbol{J}}_2\right)}{d\boldsymbol{q}} = \frac{1}{\mu(\boldsymbol{J}_1)}\left(\frac{d\mu(\boldsymbol{J})}{d\boldsymbol{q}} - \frac{d\mu(\boldsymbol{J}_1)}{d\boldsymbol{q}}\mu\left(\hat{\boldsymbol{J}}_2\right)\right) \tag{3.70}$$

The only unknown in Eq. (3.70) is the derivative of the measure of manipulability of the augmented Jacobian $\frac{d\mu(\boldsymbol{J})}{d\boldsymbol{q}}$, which can be easily computed in closed symbolic form using again the Eq. (3.37).

An alternative approach to algorithmic singularities avoidance, in the case of two tasks with order of priority, has been presented in [22] and tested on the SAUVIM manipulator. However, despite the results were comparable, Eq. (3.56) allows an easier generalization in case of more than 2 subtasks, as shown in the Sect. 3.3.3.

3.3.3 Generalization of Task Reconstruction to Multiple Subtasks

The general case of inverse kinematics in presence of multiple tasks has been addressed in Eq. (3.27):

$$\delta\boldsymbol{q} = \sum_{i=1}^{k} \hat{\boldsymbol{J}}_i^+ \delta\hat{\boldsymbol{r}}_i \tag{3.71}$$

where k the number of subtasks and $\hat{\boldsymbol{J}}_i$, $\delta\hat{\boldsymbol{r}}_i$ defined by the recursion:

$$\begin{cases} \delta\boldsymbol{q}_i = \delta\boldsymbol{q}_{i-1} + \hat{\boldsymbol{J}}_i^+ \delta\hat{\boldsymbol{r}}_i \\ \delta\hat{\boldsymbol{r}}_i = \delta\boldsymbol{r}_i - \boldsymbol{J}_i\delta\boldsymbol{q}_{i-1} \\ \boldsymbol{N}_i = \boldsymbol{N}_{i-1} - \hat{\boldsymbol{J}}_i^+\hat{\boldsymbol{J}}_i \\ \hat{\boldsymbol{J}}_i = \boldsymbol{J}_i\boldsymbol{N}_{i-1} \end{cases}, \quad \begin{cases} \delta\boldsymbol{q}_0 = \boldsymbol{0} \\ \boldsymbol{J}_0 = \boldsymbol{0} \\ \boldsymbol{N}_0 = \boldsymbol{I}_n \end{cases}, \tag{3.72}$$

The task reconstruction in case of multiple tasks prevents *every* matrix $\hat{\boldsymbol{J}}_i$ in Eq. (3.72) from losing rank. We apply the *TR* process (3.52) to the modified i-th task $\delta\hat{\boldsymbol{r}}_i$ in order to prevent the associated $\hat{\boldsymbol{J}}_i$ from losing rank. Similarly to Sect. 3.3.2 this is done after applying the *TR* process (3.52) to the previous (highest priority) tasks. Thus we simply have:

$$\delta\boldsymbol{q} = \sum_{i=1}^{k} \hat{\boldsymbol{J}}_i^+ \delta\hat{\boldsymbol{r}}_{ip} \tag{3.73}$$

with:

$$\begin{cases} \delta\boldsymbol{q}_i = \delta\boldsymbol{q}_{i-1} + \hat{\boldsymbol{J}}_i^+ \delta\hat{\boldsymbol{r}}_{ip} \\ \delta\hat{\boldsymbol{r}}_i = \delta\boldsymbol{r}_i - \boldsymbol{J}_i\delta\boldsymbol{q}_{i-1} \\ \boldsymbol{N}_i = \boldsymbol{N}_{i-1} - \hat{\boldsymbol{J}}_i^+\hat{\boldsymbol{J}}_i \\ \hat{\boldsymbol{J}}_i = \boldsymbol{J}_i\boldsymbol{N}_{i-1} \end{cases}, \quad \begin{cases} \delta\boldsymbol{q}_0 = \boldsymbol{0} \\ \boldsymbol{J}_0 = \boldsymbol{0} \\ \boldsymbol{N}_0 = \boldsymbol{I}_n \end{cases}, \tag{3.74}$$

and:

$$\begin{cases} \delta\hat{\boldsymbol{r}}_{ip} = TR\left(\hat{\boldsymbol{J}}_i, m_i(\boldsymbol{q}), \delta\hat{\boldsymbol{r}}_i\right) \\ m_i(\boldsymbol{q}) = \mu\left(\hat{\boldsymbol{J}}_i\right) = \sqrt{\det\left(\hat{\boldsymbol{J}}_i\hat{\boldsymbol{J}}_i^T\right)} \end{cases} \tag{3.75}$$

In the above formulation, the computation of $\frac{\partial m_i(\boldsymbol{q})}{\partial \boldsymbol{q}}$ is not immediate. However, it is possible to show that:

$$\prod_{i=1}^{j}\sqrt{\det\left(\hat{\boldsymbol{J}}_i\hat{\boldsymbol{J}}_i^T\right)} = \sqrt{\det\left(\boldsymbol{S}_j\boldsymbol{S}_j{}^T\right)}, \quad j = 1\ldots k. \tag{3.76}$$

where

$$\boldsymbol{S}_j = \begin{bmatrix} \boldsymbol{J}_1 \\ \boldsymbol{J}_2 \\ \vdots \\ \boldsymbol{J}_j \end{bmatrix} \tag{3.77}$$

is the matrix obtained by stacking the Jacobians $\boldsymbol{J}_i$, $i = 1\ldots j$.

Equation (3.76) allows computing the derivative of $m_i(\boldsymbol{q})$ of Eq. (3.75) with respect $\boldsymbol{q}$, using a procedure similar to the two tasks case (Sect. 3.3.2). Let's write the products of Eq. (3.76) in a different form:

$$\mu\left(\hat{\boldsymbol{J}}_j\right)\prod_{i=1}^{j-1}\mu\left(\hat{\boldsymbol{J}}_i\right) = \mu\left(\boldsymbol{S}_j\right), \quad j = 1\ldots k \tag{3.78}$$

$$\mu\left(\hat{\boldsymbol{J}}_j\right)\mu\left(\boldsymbol{S}_{j-1}\right) = \mu\left(\boldsymbol{S}_j\right), \quad j = 1\ldots k. \tag{3.79}$$

Deriving with respect to the configuration vector $\boldsymbol{q}$ we have:

$$\frac{d\mu\left(\hat{\boldsymbol{J}}_j\right)}{d\boldsymbol{q}} = \frac{1}{\mu\left(\boldsymbol{S}_{j-1}\right)}\left(\frac{d\mu\left(\boldsymbol{S}_j\right)}{d\boldsymbol{q}} - \frac{d\mu\left(\boldsymbol{S}_{j-1}\right)}{d\boldsymbol{q}}\mu\left(\hat{\boldsymbol{J}}_j\right)\right) \tag{3.80}$$

The derivative of the measure of manipulability of the augmented Jacobian $\frac{d\mu(\boldsymbol{S}_j)}{d\boldsymbol{q}}$ can be easily computed in closed symbolic form using again Eq. (3.37).

This solution allows avoiding all kinds of singularities (kinematic and algorithmic) in task-priority based kinematic controllers. The main advantage is a better tracking error in proximity of singular configurations, with respect to the order of priority of the tasks. Because the rank of the different Jacobian matrices is never zero, the presented algorithm uses the exact pseudo-inversion and the resulting task errors depend only on the choice of the lower limit of the distance from singularity.

This approach is suitable for autonomous systems because it ensures avoiding every kind of singularity regardless of the input task. If, for example, the task planner requires the arm to reach a particular configuration close to or in a singular point, the manipulator will execute the task with an error as small as possible, avoiding the singular point. The path planner could be informed of this error and then would make an appropriate action.

3.3.4 Experimental Results

This section presents some experimental results of the *TR* algorithm, used in the control system for the underwater manipulator of SAUVIM. The experiments were conducted for the following cases:

A. Single task with kinematic singularity
B. Two tasks with algorithmic singularity

In any of the above cases, the results with the task reconstruction algorithm were compared with the results obtained using the damped least-squares inverse with numerical filtering (DLS method, [7]) which is widely used for avoiding the kinematic singularity. Its formulation is:

$$\boldsymbol{J}^{+}_{DLS} = \boldsymbol{J}^T \left(\boldsymbol{J}\boldsymbol{J}^T + \lambda^2 \boldsymbol{u}_m \boldsymbol{u}_m^T\right)^{-1}, \tag{3.81}$$

$$\lambda^2 = \begin{cases} 0, & \sigma_m \geq \epsilon \\ \left(1 - \left(\frac{\sigma_m}{\epsilon}\right)^2\right) \lambda^2_{max}, & \sigma_m < \epsilon \end{cases} \tag{3.82}$$

where σ_m is the lowest singular value of $\boldsymbol{J}$ and $\boldsymbol{u}_m$ is the corresponding output singular vector.

3.3.4.1 MARIS 7080 Manipulator

The following experiments have been conducted with the SAUVIM manipulator MARIS 7080 (Fig. 3.7), already introduced in Sect. 1.2. Some details about MARIS 7080 have been presented also in [23–26].

3.3.4.2 Case I: Single Task with Kinematic Singularity

For the first experiment the desired task is a circular trajectory on the $X - Z$ plane, with a radius of 0.3 m, centered in:

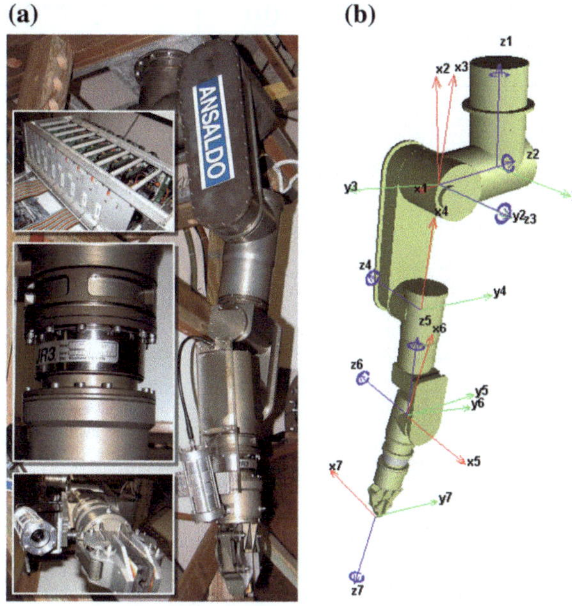

Fig. 3.7 The 7-DOF SAUVIM manipulator MARIS 7080 and its kinematics

$$\boldsymbol{x}_0 = \begin{bmatrix} 0.5 \\ 0 \\ -1 \end{bmatrix}. \tag{3.83}$$

The manipulation variable in Eq. (3.51) is the position $\boldsymbol{x}(t) \in \Re^3$. With this configuration, a part of the desired task lies outside of the workspace: this causes the manipulator to encounter a kinematic singularity. The parameters are set as follows:

- TR Method (Eq. 3.52): $\overline{m} = 0.05$, $K_r = 0.5$.
- DLS Method (Eq. 3.81): $\epsilon = 0.2$, $\lambda_{max} = 0.1$

For the closed-loop controller of Fig. (3.2), the controller gain is $\gamma = 10$, tuned for the best task performance and stability.

The results are shown in Fig. 3.8. With the *TR* method, the singularity measure is clamped over $\overline{m} = 0.05$ (Fig. 3.8b), while using the *DLS* the measure goes close to zero (Fig. 3.8d). In this situation, a high gain or a large task error may easily result in instability of the system.

Figure 3.8e shows finally the task error comparison. While the *TR* algorithm has a clean and predictable behavior, the DLS method shows an overshoot upon exiting from the singular region. This is due to the errors introduced by the damping term λ of Eq. (3.81).

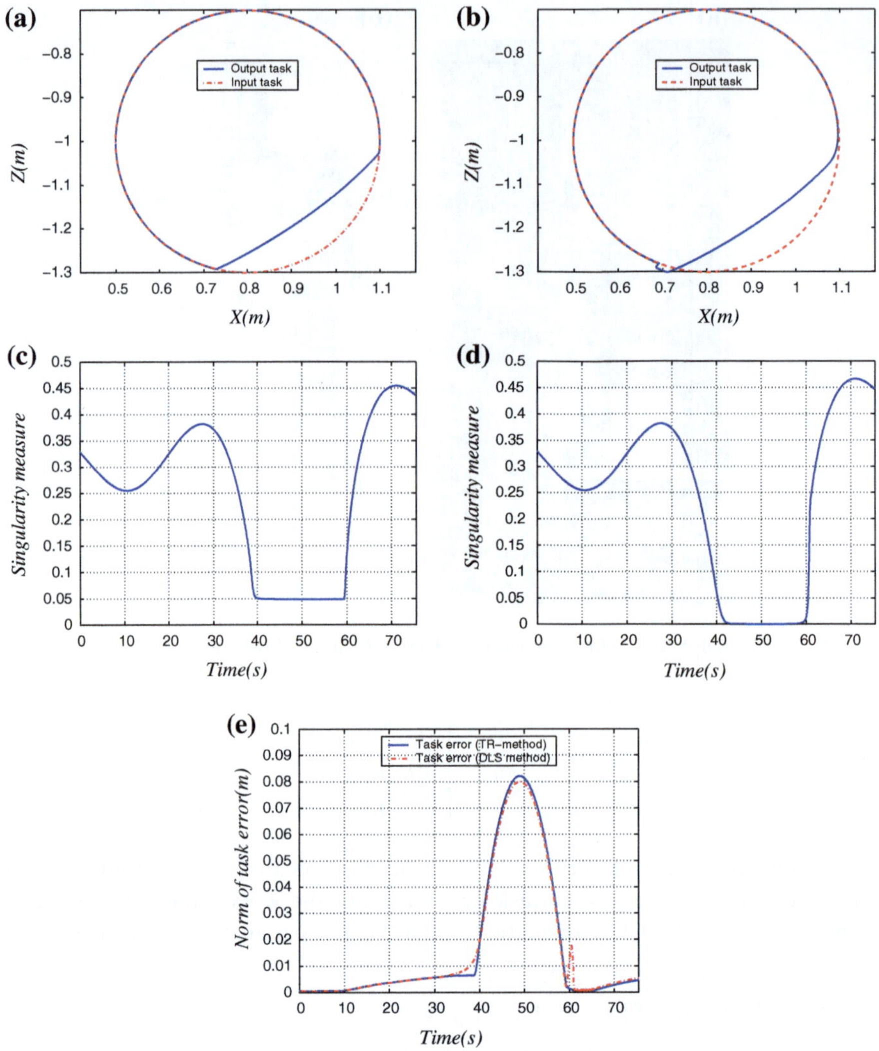

Fig. 3.8 Experimental results (Case I): **a** task in the x-z plane (TR-method), **b** task in the x-z plane (DLS method), **c** singularity measure (TR-method), **d** singularity measure (DLS method), **e** norm of task errors

3.3.4.3 Case II: Two Tasks with Algorithmic Singularity

For the second experiment, we have two subtasks (position and orientation) with the order of priority (position and then orientation). The desired first task (position) is a circular trajectory on the $X - Z$ plane, with a radius of 0.2 m, centered in:

$$\boldsymbol{x}_0 = \begin{bmatrix} 0.55 \\ 0 \\ -1 \end{bmatrix}. \tag{3.84}$$

The secondary task is simply a constant RPY angle, $(\pi, 0, -\pi/2)$. In this case, the *TR* algorithm is given by Eq. (3.56). With this configuration, there is no kinematic singularity associated to the primary task. However, some part of the desired secondary task is conflicting with the primary: this causes the manipulator to encounter an algorithmic singularity.

The results are compared with those of the Chiaverini's algorithmic singularity avoidance method [7],

$$\delta \boldsymbol{q} = \boldsymbol{J_1}^{+} \delta \boldsymbol{r}_1 + \left(\boldsymbol{I} - \boldsymbol{J}_1^{+} \boldsymbol{J}_1\right) \boldsymbol{J}_2^{+} \delta \boldsymbol{r}_2.. \tag{3.85}$$

The pseudo-inversions of $\boldsymbol{J}_1$ and $\hat{\boldsymbol{J}}_2$ are still performed using Eq. (3.81). The parameters are set as follows:

- TR Method (Eq. 3.56): $\overline{m}_1 = 0.05, \overline{m}_2 = 0.075, K_r1 = K_r2 = 0.5.$
- DLS Method (Eq. 3.81): $\epsilon = 0.2$, $\lambda_{max} = 0.1$

The task controller gains are the same as used in Case I, for both methods.

Figure 3.9 shows the results of this case. The primary tasks are well performed in both methods. Instead, for the secondary task, with the TR-method (Fig. 3.9c) the singularity measure is maintained above 0.075 as our desire. Therefore, there are no secondary task singularities.

Figure 3.9e shows the comparison of the norm of the secondary task errors. The larger error present when using Chiaverini's method is intrinsic in the nature of Eq. (3.85). As noted in [7], when an algorithmic singularity is approached, the solution of Eq. (3.85) progressively reduces the null-space velocity associated to the component of the secondary task that is close to become infeasible. The TR method, conversely, does not introduce any changes into the original task solution (3.21) outside of the threshold area. When an algorithmic singularity is approached, the above solution of Eq. (3.21) increases the null-space velocity to preserve accurate tracking of the secondary task close to become unfeasible. Just before approaching the singularity, the task reconstruction intervenes, modifying the trajectory in order to maintain the singularity measure above the predefined threshold, as shown in Fig. 3.9c.

This aspect also justifies why there is an inversion of behaviors in the configuration close to the center of the error peak in Fig. 3.9e. Here, the Chiaverini method tends to asymptotically reduce the error, even if with slower dynamics, pushing the overall configuration closer to the singularity. With the TR method, this is not allowed, and the larger error is necessary in order to maintain the distance from singularity over the predefined threshold.

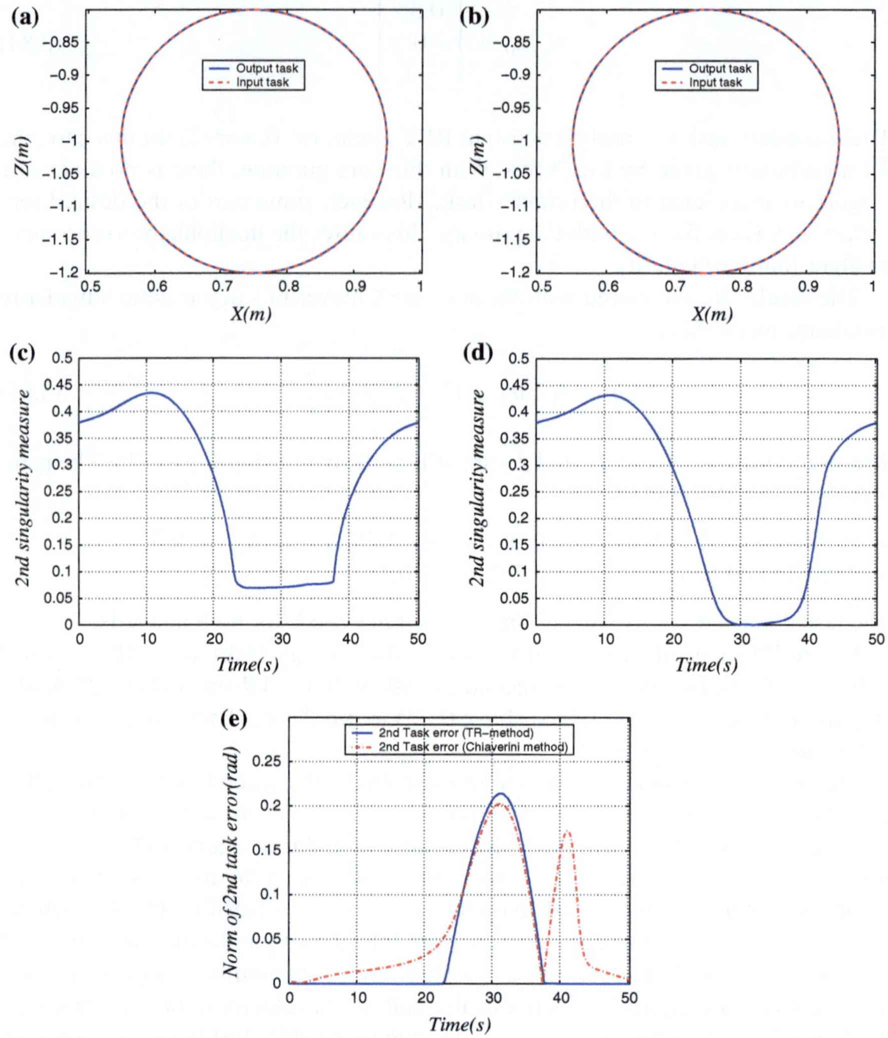

Fig. 3.9 Experimental results (Case II): **a** task in the *x*-*z* plane (TR-method), **b** task in the *x*-*z* plane (Chiaverini), **c** 2nd task singularity measure (TR-method), **d** 2nd task singularity measure (Chiaverini), **e** norm of the 2nd task errors

In conclusion, the TR-method can prevent the algorithmic singularity, with a clean behavior and a smaller error in the area close to the singularity. The predictability of its behavior makes the Task Reconstruction algorithm more suitable for autonomous manipulation.

References

1. Sciavicco L, Siciliano B (2001) Modeling and control of robot manipulators, 2nd edn. Springer, Berlin
2. Aicardi M, Caiti A, Cannata G, Casalino G (1995) Stability and robustness analysis of a two layered hierarchical architecture for the closed loop control of robots in the operational space. In: Proceedings of 1995 IEEE international conference on robotics and automation, 1995, vol 3, pp 2771–2778. doi:10.1109/ROBOT.1995.526005
3. Whitney DE (1969), Resolved motion rate control of manipualtors and human prostheses. IEEE Trans Man-Mach Syst MMS-10(2):47–53
4. Nakamura Y (1991) Advanced robotics: redundancy and optimization. Addison Wesley, Reading
5. Penrose R (1955) A generalized inverse for matrices. Math Proc Cambridge Philos Soc 51(03):406–413
6. Wampler CW (1986) Manipulator inverse kinematic solutions based on vector formualtions and damped least-squares methods. IEEE Trans Syst Man Cybern SMC-16(1):93–101
7. Chiaverini S (1997) Singularity-robust task-priority redundancy resolution for real-time kinematic control of robot manipulators. IEEE Trans Robot Autom 13(3):398–410
8. Nakamura Y, Hanafusa H (1986) Inverse kinematic solutions with singularity robustness for robot manipulator control. J Dyanmic Syst Meas Control 108:163–171
9. Siciliano B, Slotine JJE (1991) A general framework for managing multiple tasks in highly redundant robotic systems. In: Proceedings of international conference on advanced robotics, pp 1211–1216
10. Baerlocher P, Boulic R (1998) Task-priority formulations for the kinematic control of highly redundant articulated structures. In: Proceedings of IEEE/RSJ international conference on intelligent robots and systems, pp 323–329
11. Maciejewski AA, Klein CA (1985) Obstacle avoidance for kinematically redunadant manipulators in dynamically varying environments. Int J Robot Res 4(3):109–117
12. Park J, Choi Y, Chung WK, Youm Y (2001) Multiple tasks kinematics using weighted pseudo-inverse for kinematically redundant manipulators. In: Proceedings of 2001 ICRA. IEEE international conference on robotics and automation, vol 4, pp 4041–4047. doi:10.1109/ROBOT.2001.933249
13. Kim J, Marani G, Chung WK, Yuh J (2006) Task reconstruction method for real-time singularity avoidance for robotic manipulators. Adv Robot 20(4):453–481
14. Marani G, Kim J, Chung WK, Yuh J (2003) Algorithmic singularities avoidance in task-priority based controller for redundant manipulators. In: Proceedings of IEEE/RSJ international conference on intelligent robots and systems, pp 1942–1947
15. Marani G, Kim J, Yuh J, Chung WK (2002) A real-time approach for singularity avoidance in resolved motion rate control of robotic manipulators. In: Proceedings of IEEE international conference on robotics and automation, pp 1973–1978
16. Kim J, Marani G, Chung WK, Yuh J, Oh SR (2002b) Dynamic task priority approach to avoid kinematic singularity for autonomous manipulation. In: Proceedings of IEEE/RSJ international conference on intelligent robots and systems, pp 1942–1947
17. Yoshikawa T (1985) Manipulability of robotic mechanisms. Int J Robot Res 4(2):3–9
18. Park J (1999) Analysis and control of kinematically redundant manipulators: an approach based on kinematically decoupled joint space decomposition. Ph.D. thesis, Pohang University of Science and Technology (POSTECH)
19. Kim J, Marani G, Chung WK, Yuh J (2002) Kinematic singularity avoidance for autonomous manipulation in underwater. In: The fifth ISOPE Pacific/Asia offshore mechanics symposium, Daejeon, Korea
20. Kim J, Marani G, Chung WK, Yuh J (2004) A general singularity avoidance framework for robot manipulators: task reconstruction method. In: ICRA, New Orleans, USA, pp 4809–4814

21. Seraji H, Colbaugh R (1990) Singularity-robustness and task-prioritization in configuration control of redundant robots. In: Proceedings of IEEE conference on decision and control, pp 3089–3095
22. Marani G, Yuh J, Choi SK (2006) Autonomous manipulation for an intervention auv. In: Sutton B, Roberts G (eds) Guidance and control of unmanned marine vehicles. IEE's control engineering series, pp 217–237
23. Marani G, Bozzo T, Choi SK, (2000) A fast prototyping approach for designing the maris manipulator control. In: Symposium on underwater robotic technology (SURT 2000), Wailea, Maui, Hawaii
24. Yuh J, Marani G (2001) An advanced underwater robotic manipulator for SAUVIM. In: 2001 IEEE international conference on robotics and automation, workshop W5 (Underwater Robotic Technologies), Seoul, Korea
25. Yuh J, Choi S, Kim T, Marani G, West M, Easterday O, Rosa K (2003) Real-time control architecture for SAUVIM. In: 1st IFAC workshop on guidance and control of underwater vehicles, Newport, South Wales, UK
26. Marani G, Medrano I, Choi SK, Yuh J (2005) A client-server oriented programming language for autonomous underwater manipulation. In: The proceedings of the fifteenth international offshore and polar engineering conference, Seoul, Korea

Chapter 4
The SAUVIM Underwater Vehicle-Manipulator System

In the previous chapters we presented a methodology for describing generalized robotic structures, including their representation (Sect. 2.1.3), forward kinematics computation (Sect. 2.2), full system dynamics (Sect. 2.3) and resolution of inverse kinematics (Chap. 3).

The inverse kinematic problem will now be applied to the entire SAUVIM vehicle, modeled as a 13 degrees-of-freedom multi-body system by using the previous results.

By applying the Task Reconstruction to the whole structure, we will show how it is possible to autonomously optimize the placement of the vehicle and arm with respect to the target. It is often regarded in literature as workspace optimization.

4.1 Modeling the SAUVIM Vehicle-Manipulator System

SAUVIM, as introduced in Sect. 1.2 (Fig. 1.1), was built around an open-frame structure, enclosed by a flooded composite fairing. Navigation and hovering movements are precisely actuated with eight thrusters located around the center of mass.

To achieve the intervention capabilities, SAUVIM is equipped with a seven degrees-of-freedom robotic manipulator, MARIS 7080 (Fig. 3.7, see Sect. 1.2).

With the considerations introduced in Sect. 2.1.3, this platform can be regarded as a linear chain of 8 joints, with the first joint having 6 DOF. Figure 4.1 schematizes this concept. As usual, we will be numbering the frame sequence from 1 to 8, indicating with $\langle 0 \rangle$ the base frame. Note that our choice was to make the link-1 frame coincident with the center of mass frame, as shown in Fig. 4.2. Instead, the link-2 frame of the underwater vehicle-manipulator system (UVMS) coincides with the link-1 frame of the manipulator (see Fig. 4.3).

By following the procedure of Sect. 2.2.2, each joint is being represented by an associated joint matrix, listed in Tables 4.1 and 4.2.

The configuration vector of the structure is hence defined as:

$$\boldsymbol{q} = \begin{bmatrix} q_1 \; q_2 \; \dots \; q_{13} \end{bmatrix}^T \tag{4.1}$$

G. Marani and J. Yuh, *Introduction to Autonomous Manipulation*,
Springer Tracts in Advanced Robotics 102, DOI: 10.1007/978-3-642-54613-6_4,

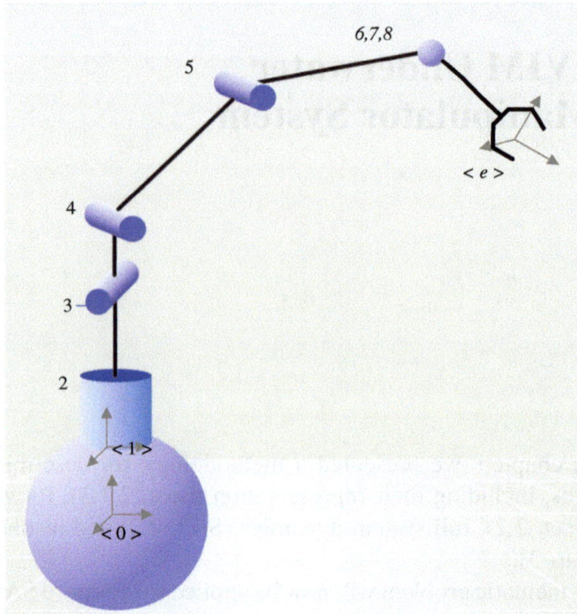

Fig. 4.1 Model schematization of the SAUVIM vehicle-manipulator system

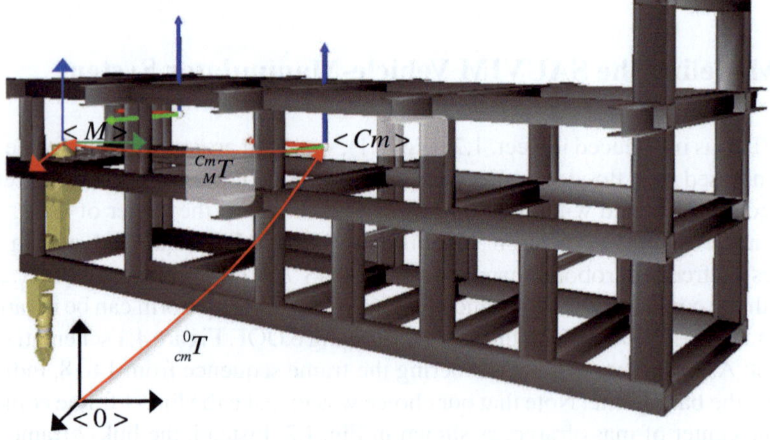

Fig. 4.2 Reference systems distribution on vehicle

and the quasivelocity vector is:

$$\boldsymbol{p} = \left[p_1 \; p_2 \; \ldots \; p_{13} \right]^T \tag{4.2}$$

The physical meanings of $\boldsymbol{q}$ and $\boldsymbol{p}$ will be more clear in the following paragraphs.

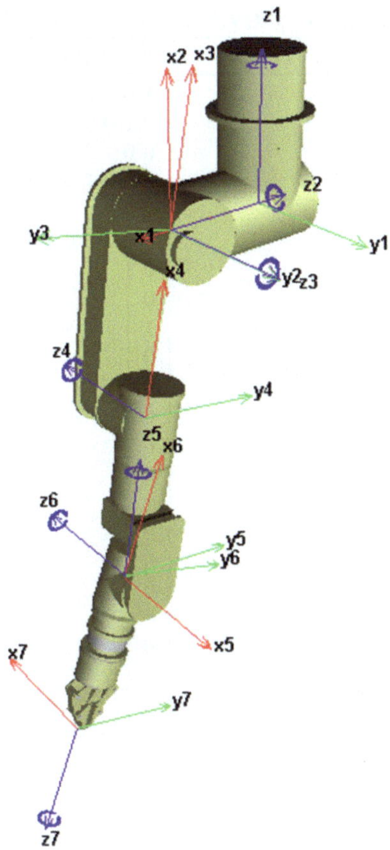

Fig. 4.3 Kinematics of the 7 DOF SAUVIM manipulator MARIS 7080

Joint 1

The first kinematic linkage of the system is modeled as a 6 DOF simple joint. Its joint matrix is listed in Table 4.1 and is simply the 6×6 identity matrix. This joint matrix simply represents a free body, i.e. the link is capable of performing any rotation and translation in the Cartesian space.

By following the procedure indicated in Sect. 2.2.2.1 we obtain the transformation matrix:

$$
{}^{0}_{1}\boldsymbol{T} = \begin{bmatrix} \cos(q3)\cos(q2) & -\sin(q3)\cos(q1) + \cos(q3)\sin(q2)\sin(q1) & \sin(q3)\sin(q1) + \cos(q3)\sin(q2)\cos(q1) & q4 \\ \sin(q3)\cos(q2) & \cos(q3)\cos(q1) + \sin(q3)\sin(q2)\sin(q1) & -\cos(q3)\sin(q1) + \sin(q3)\sin(q2)\cos(q1) & q5 \\ -\sin(q2) & \cos(q2)\sin(q1) & \cos(q2)\cos(q1) & q6 \\ 0 & 0 & 0 & 1 \end{bmatrix} \quad (4.3)
$$

In Eq. (4.3) the configuration variables q_1 through q_6 represent respectively the three Euler angles (roll, pitch and yaw) and the Cartesian coordinates of the origin of the frame $\langle 1 \rangle$, located at the center of mass of the vehicle (see Fig. 4.2).

Computing the Jacobian of the origin of frame $\langle 1 \rangle$, with the procedure in Sect. 2.2.3, gives the following result:

Table 4.1 SAUVIM UVMS description (joints 1–4)

Joint number	Joint matrix	Outboard joint	Local frame rotation
Joint 1	$\begin{bmatrix} 1&0&0&0&0&0\\0&1&0&0&0&0\\0&0&1&0&0&0\\0&0&0&1&0&0\\0&0&0&0&1&0\\0&0&0&0&0&1 \end{bmatrix}$	$\begin{bmatrix} x_{M1}\\y_{M1}\\z_{M1} \end{bmatrix}$	$\begin{bmatrix} 1&0&0\\0&1&0\\0&0&1 \end{bmatrix}$
Joint 2	$\begin{bmatrix} 0\\0\\1\\0\\0\\0 \end{bmatrix}$	$\begin{bmatrix} 0.2235\\0\\0 \end{bmatrix}$	$\begin{bmatrix} 1&0&0\\0&1&0\\0&0&1 \end{bmatrix}$
Joint 3	$\begin{bmatrix} 0\\0\\1\\0\\0\\0 \end{bmatrix}$	$\begin{bmatrix} 0\\0\\0 \end{bmatrix}$	$\begin{bmatrix} 0&0&-1\\0&1&0\\1&0&0 \end{bmatrix}$
Joint 4	$\begin{bmatrix} 0\\0\\1\\0\\0\\0 \end{bmatrix}$	$\begin{bmatrix} -0.4\\0\\0 \end{bmatrix}$	$\begin{bmatrix} 1&0&0\\0&0&1\\0&-1&0 \end{bmatrix}$

$$ {}^{1}J_{1/0}\,(O_1) = \begin{bmatrix} 1&0&0&0&0&0&0&0&0&0&0&0&0&0\\ 0&1&0&0&0&0&0&0&0&0&0&0&0&0\\ 0&0&1&0&0&0&0&0&0&0&0&0&0&0\\ 0&0&0&1&0&0&0&0&0&0&0&0&0&0\\ 0&0&0&0&1&0&0&0&0&0&0&0&0&0\\ 0&0&0&0&0&1&0&0&0&0&0&0&0&0 \end{bmatrix} \tag{4.4} $$

As noted in Eq. (2.154), the above Jacobian figures in the computation of the generalized velocity of the origin of the frame $\langle 1 \rangle$ projected in the same frame $\langle 1 \rangle$:

$$ {}^{1}\boldsymbol{X}_{O_1/0} = {}^{1}\boldsymbol{J}_{1/0}\,(O_1)\boldsymbol{p} = \begin{bmatrix} p_1\\p_2\\p_3\\p_4\\p_5\\p_6 \end{bmatrix} \tag{4.5} $$

Table 4.2 SAUVIM UVMS description (joints 5–8)

Joint number	Joint matrix	Outboard joint	Local frame rotation
Joint 5	$\begin{bmatrix}0\\0\\1\\0\\0\\0\end{bmatrix}$	$\begin{bmatrix}-0.4\\0\\0\end{bmatrix}$	$\begin{bmatrix}1&0&0\\0&-1&0\\0&0&-1\end{bmatrix}$
Joint 6	$\begin{bmatrix}0\\0\\1\\0\\0\\0\end{bmatrix}$	$\begin{bmatrix}0\\0\\0\end{bmatrix}$	$\begin{bmatrix}0&0&1\\0&1&0\\-1&0&0\end{bmatrix}$
Joint 7	$\begin{bmatrix}0\\0\\1\\0\\0\\0\end{bmatrix}$	$\begin{bmatrix}0\\0\\0\end{bmatrix}$	$\begin{bmatrix}0&0&-1\\0&1&0\\1&0&0\end{bmatrix}$
Joint 8	$\begin{bmatrix}0\\0\\1\\0\\0\\0\end{bmatrix}$	$\begin{bmatrix}0\\0\\0.414\end{bmatrix}$	$\begin{bmatrix}0&0&-1\\0&1&0\\1&0&0\end{bmatrix}$

Equation (4.5) shows that the quasi-velocities p_1, p_2, p_3 represent the angular velocities around the three coordinate axis $\boldsymbol{i}, \boldsymbol{j}$ and $\boldsymbol{k}$ respectively, while p_4, p_5 and p_6 are the Cartesian velocities (all projected in the frame $\langle 1 \rangle$).

Note that the coordinates of the outboard joint in Table 4.1 are simply the geometric vector connecting the origin of the first joint frame of the Maris manipulator to the center of mass of the vehicle. The origin of the first joint frame is shown in Fig. 4.3.

Joint 2 through 8

Joints 2 through 8 of the UVMS correspond to the joints 1 through 7 of the MARIS 7080 manipulator. For consistency with the Denavit-Hartenberg representation, the local link frames have been oriented so that the z axis coincides with the rotational motion axis. This results in the same joint matrix for all the manipulator joints, as shown in Tables 4.1 and 4.2. Note that the configuration variables q_7 through q_{13} represent the joint angles of the robot, while the quasi-velocities variables p_7 through p_{13} are the joint angular velocities.

By following again the procedure in Sect. 2.2.2.1 we obtain the following transformation matrices:

$$ {}^{1}_{2}\boldsymbol{T} = \begin{bmatrix} \cos(q_7) & -\sin(q_7) & 0 & x_{M1} \\ \sin(q_7) & \cos(q_7) & 0 & y_{M1} \\ 0 & 0 & 1 & z_{M1} \\ 0 & 0 & 0 & 1 \end{bmatrix} \tag{4.6} $$

$$ {}^{2}_{3}\boldsymbol{T} = \begin{bmatrix} 0.0 & 0.0 & -1 & 0.2235 \\ \sin(q_8) & \cos(q_8) & 0 & 0 \\ \cos(q_8) & -\sin(q_8) & 0 & 0 \\ 0 & 0 & 0 & 1 \end{bmatrix} \tag{4.7} $$

$$ {}^{3}_{4}\boldsymbol{T} = \begin{bmatrix} \cos(q_9) & -\sin(q_9) & 0 & 0 \\ 0 & 0 & 1 & 0 \\ -\sin(q_9) & -\cos(q_9) & 0 & 0 \\ 0 & 0 & 0 & 1 \end{bmatrix} \tag{4.8} $$

$$ {}^{4}_{5}\boldsymbol{T} = \begin{bmatrix} \cos(q_{10}) & -\sin(q_{10}) & 0 & -0.4 \\ -\sin(q_{10}) & -\cos(q_{10}) & 0 & 0 \\ 0 & 0 & -1 & 0 \\ 0 & 0 & 0 & 1 \end{bmatrix} \tag{4.9} $$

$$ {}^{5}_{6}\boldsymbol{T} = \begin{bmatrix} 0 & 0 & 1 & -0.4 \\ \sin(q_{11}) & \cos(q_{11}) & 0 & 0 \\ -\cos(q_{11}) & \sin(q_{11}) & 0 & 0 \\ 0 & 0 & 0 & 1 \end{bmatrix} \tag{4.10} $$

$$ {}^{6}_{7}\boldsymbol{T} = \begin{bmatrix} 0 & 0 & -1 & 0 \\ \sin(q_{12}) & \cos(q_{12}) & 0 & 0 \\ \cos(q_{12}) & -\sin(q_{12}) & 0 & 0 \\ 0 & 0 & 0 & 1 \end{bmatrix} \tag{4.11} $$

$$ {}^{7}_{8}\boldsymbol{T} = \begin{bmatrix} 0 & 0 & -1 & 0 \\ \sin(q_{13}) & \cos(q_{13}) & 0 & 0 \\ \cos(q_{13}) & -\sin(q_{13}) & 0 & 0 \\ 0 & 0 & 0 & 1 \end{bmatrix} \tag{4.12} $$

Similarly, after applying the procedure in Sect. 2.2.3, the computation of the Jacobian of the origin of frames 2 through 8 gives the following results:

$$ {}^{2}\boldsymbol{J}_{2/0}(O_2) = \begin{bmatrix} \cos(q_7) & \sin(q_7) & 0 & 0 & 0 & 0\,0\,0\,0\,0\,0\,0\,0 \\ -\sin(q_7) & \cos(q_7) & 0 & 0 & 0 & 0\,0\,0\,0\,0\,0\,0\,0 \\ 0 & 0 & 1 & 0 & 0 & 0\,1\,0\,0\,0\,0\,0\,0 \\ -\sin(q_7)z_{M1} & \cos(q_7)z_{M1} & -\cos(q_7)y_{M1}+\sin(q_7)x_{M1} & \cos(q_7) & \sin(q_7) & 0\,0\,0\,0\,0\,0\,0\,0 \\ -\cos(q_7)z_{M1} & -\sin(q_7)z_{M1} & \sin(q_7)y_{M1}+\cos(q_7)x_{M1} & -\sin(q_7) & \cos(q_7) & 0\,0\,0\,0\,0\,0\,0\,0 \\ y_{M1} & -x_{M1} & 0 & 0 & 0 & 1\,0\,0\,0\,0\,0\,0\,0 \end{bmatrix} \tag{4.13} $$

$$
\begin{aligned}
&{}^{3}\boldsymbol{J}_{3/0}(O_3) \\
&= \left[\begin{array}{c}
-\sin(q8)\sin(q7) \\
-\cos(q8)\sin(q7) \\
-1.0\cos(q7) \\
0.2235\cos(q8)\sin(q7) - \sin(q8)\cos(q7)\, z_{MI} + \cos(q8)\, y_{MI} \\
-0.2235\sin(q8)\sin(q7) - \cos(q8)\cos(q7)\, z_{MI} - \sin(q8)\, y_{MI} \\
1.0\sin(q7)\, z_{MI}
\end{array}\right. \\
&\times \begin{array}{c}
\sin(q8)\cos(q7) \\
\cos(q8)\cos(q7) \\
-1.0\sin(q7) \\
-0.2235\cos(q8)\cos(q7) - \sin(q8)\sin(q7)\, z_{MI} - \cos(q8)\, x_{MI} \\
0.2235\sin(q8)\cos(q7) - \cos(q8)\sin(q7)\, z_{MI} + \sin(q8)\, x_{MI} \\
-1.0\cos(q7)\, z_{MI}
\end{array} \\
&\times \begin{array}{c}
\cos(q8) \\
-\sin(q8) \\
-0.0 \\
0.2235\sin(q8) + \sin(q8)(\sin(q7)\, y_{MI} + \cos(q7)\, x_{MI}) \\
0.2235\cos(q8) + \cos(q8)(\sin(q7)\, y_{MI} + \cos(q7)\, x_{MI}) \\
1.0\cos(q7)\, y_{MI} - 1.0\sin(q7)\, x_{MI}
\end{array} \\
&\times \left.\begin{array}{cccccccccc}
0.0 & 0.0 & 0.0 & \cos(q8) & 0 & 0 & 0 & 0 & 0 & 0 \\
0.0 & 0.0 & 0.0 & -\sin(q8) & 0 & 0 & 0 & 0 & 0 & 0 \\
-0.0 & -0.0 & -0.0 & -0.0 & 1 & 0 & 0 & 0 & 0 & 0 \\
-\sin(q8)\sin(q7) & \sin(q8)\cos(q7) & \cos(q8) & 0.2235\sin(q8) & 0 & 0 & 0 & 0 & 0 & 0 \\
-\cos(q8)\sin(q7) & \cos(q8)\cos(q7) & -\sin(q8) & 0.2235\cos(q8) & 0 & 0 & 0 & 0 & 0 & 0 \\
-1.0\cos(q7) & -1.0\sin(q7) & 0.0 & 0.0 & 0 & 0 & 0 & 0 & 0 & 0
\end{array}\right]
\end{aligned}
\tag{4.14}
$$

The increased complexity of the successive Jacobians does not allow to show them here.

As a matter of fact, the description introduced in Chap. 2 has been formulated for simplifying the automated generation of the equations from the description. The entire process has been implemented as a set of procedures within a symbolic-math processor software. With the only input of the data of Tables 4.1 and 4.2, the symbolic model generator provides all the kinematic quantities optimized for real-time implementation. Large problems like the present 13 DOF model of the UVMS would be extremely complex to solve without such an automated procedure.

4.2 Workspace Optimization with Task Reconstruction

In Sect. 3.3 we have introduced the concept of task reconstruction. The TR method was applied to the MARIS 7080 manipulator to show its effectiveness in maintaining the measure of manipulability confined within reasonable values.

The TR still holds in case of multi-body systems with joints with more than one degree of freedom as, for example, the 13 DOF UVMS system presented in the previous section. The idea in workspace optimization is to apply the TR methodology to the whole 13 DOF chain, with the same goal of confining the manipulability of the manipulator within reasonable thresholds. Unlike the examples in Sect. 3.3.4, the entire structure (hence including navigation thrusters) is now considered for the optimization of the manipulability.

4.2.1 Task Formulation for Workspace Optimization

The goal of a generic intervention mission is usually achieved by performing Cartesian relative motion between the end-effector and the target. In order to avoid mission aborts the target must be optimally positioned within the dexterous arm workspace.

This concept can be better illustrated by introducing the following manipulator variables (already defined in Sect. 3.2.1):

Task $\boldsymbol{r}_1$ (3 DOF): Cartesian position of Link 8 (the end-effector)
Task $\boldsymbol{r}_2$ (3 DOF): Orientation of Link 8 (the end-effector)
Task $\boldsymbol{r}_3$ (3 DOF): Position (x and y) and orientation (around z) of the vehicle.

The order of priority is decreasing, i.e. $\boldsymbol{r}_1$ has higher priority than $\boldsymbol{r}_2$ and $\boldsymbol{r}_3$. The manipulation variables $\boldsymbol{r}_1$ and $\boldsymbol{r}_2$ must track respectively the required tool position and orientation. Instead, the manipulation variable $\boldsymbol{r}_3$ is necessary to set the hovering condition. In our implementation, the latter is achieved by setting to zero the time derivative of the third manipulation variable, i.e. of the following velocities:

- Angular velocity around z
- Linear velocity along x
- Linear velocity along y.

In fact, roll and pitch velocities are set independently, as described later, for aligning the center of buoyancy over the center of mass, and the vertical velocity is used to maintain the depth at a fixed distance from the bottom.

By applying directly the procedure in Sect. 3.3.3 to the whole 13 DOF UVMS, the interesting effect is that the vehicle position is autonomously changed only when it is needed, to increase the manipulability of the arm. In general, setting $\dot{\boldsymbol{r}}_3$ to zero means that the vehicle is stationary. However, when the arm reaches the workspace limit (due to for example slow position drifts), the position of the vehicle (task $\boldsymbol{r}_3$) is modified by the TR process in order to avoid further reduction of the manipulability.

4.2.1.1 Application Example: The SAUVIM Recovery Task

Let's examine in detail the application of the task reconstruction to the full 13 DOF SAUVIM model described in Sect. 4.1.

Let our *first manipulation variable* $\boldsymbol{r}_1 \in \Re^3$ be the cartesian position of the end-effector w.r.t. the main frame $\langle 0 \rangle$. Its time derivative, expressed as a function of the global quasivelocity vector $\boldsymbol{p}$, is:

$$\dot{\boldsymbol{r}}_1 = \boldsymbol{J}_1\boldsymbol{p} \tag{4.15}$$

where $\boldsymbol{J}_1 \in \Re^{3\times 13}$ is Jacobian of the task $\boldsymbol{r}_1$.

In SAUVIM, the end-effector frame is defined w.r.t. the last link frame $\langle 8\rangle$ by the following transformation matrix:

$$^{8}_{E}\boldsymbol{T} = \begin{bmatrix} 1\,0\,0 & 0 \\ 0\,1\,0 & 0 \\ 0\,0\,1 & 0.414 \\ 0\,0\,0 & 1 \end{bmatrix} \tag{4.16}$$

In other words, the frame $\langle E\rangle$ is obtained by simply translating the last link frame $\langle 8\rangle$ of 0.414 m along the $\boldsymbol{k}$ axis. The generalized velocity of the origin O_E of the end-effector frame $\langle E\rangle$ w.r.t. the main frame $\langle 0\rangle$ and projected on the end-effector frame $\langle E\rangle$ can be expressed according to Eq. (2.154):

$$^{E}\dot{\boldsymbol{X}}_{E/0} = {}^{E}\boldsymbol{J}_{E/0}\,(O_E)\,\boldsymbol{p} \tag{4.17}$$

with $O_E = \begin{bmatrix} 0 & 0 & 0 \end{bmatrix}^T$ and $\boldsymbol{p} \in \Re^{13}$ being the quasi-velocity vector of SAUVIM. The Jacobian $^{E}\boldsymbol{J}_{E/0}\,(O_E)$ can be easily computed from the Jacobian of frame $\langle 8\rangle$ by using Eq. (2.141):

$$^{8}\dot{\boldsymbol{X}}_{O_E/0} = \boldsymbol{\phi}\,(O_e)\,{}^{8}\dot{\boldsymbol{X}}_{E/0} = \begin{bmatrix} 1 & 0 & 0\,0\,0\,0 \\ 0 & 1 & 0\,0\,0\,0 \\ 0 & 0 & 1\,0\,0\,0 \\ 0 & 0.414 & 0\,1\,0\,0 \\ -0.414 & 0 & 0\,0\,1\,0 \\ 0 & 0 & 0\,0\,0\,1 \end{bmatrix} {}^{8}\dot{\boldsymbol{X}}_{E/0} \tag{4.18}$$

and hence:

$$^{E}\boldsymbol{J}_{E/0} = {}^{8}\boldsymbol{J}_{E/0} = \boldsymbol{\phi}\,(O_e)\,{}^{8}\boldsymbol{J}_{8/0} \tag{4.19}$$

being $^{8}_{E}\boldsymbol{R} = \boldsymbol{I}$.

The Jacobian $\boldsymbol{J}_1 \in \Re^{3\times 13}$ of the first manipulation variable $\boldsymbol{r}_1$ is then made of the last three rows of $^{E}\boldsymbol{J}_{E/0}$:

$$\boldsymbol{J}_1 = \begin{bmatrix} 0\,0\,0\,1\,0\,0 \\ 0\,0\,0\,0\,1\,0 \\ 0\,0\,0\,0\,0\,1 \end{bmatrix} {}^{E}\boldsymbol{J}_{E/0} \tag{4.20}$$

Let now our *second manipulation variable* $\boldsymbol{r}_2 \in \Re^3$ be the orientation of the end-effector w.r.t. the main frame $\langle 0\rangle$. Its time derivative, expressed as a function of the global quasivelocity vector $\boldsymbol{p}$, is:

$$\dot{\boldsymbol{r}}_2 = \boldsymbol{J}_2\boldsymbol{p} \tag{4.21}$$

where $\boldsymbol{J}_2 \in \Re^{3 \times 13}$ is Jacobian of the task $\boldsymbol{r}_2$.

Following the same considerations for the first manipulation variable, the Jacobian $\boldsymbol{J}_2 \in \Re^{3 \times 13}$ of the second manipulation variable $\boldsymbol{r}_2$ is made of the first three rows of ${}^{E}\boldsymbol{J}_{E/0}$:

$$\boldsymbol{J}_2 = \begin{bmatrix} 1\,0\,0\,0\,0\,0 \\ 0\,1\,0\,0\,0\,0 \\ 0\,0\,1\,0\,0\,0 \end{bmatrix} {}^{E}\boldsymbol{J}_{E/0} \tag{4.22}$$

Note that, as noted for Eq. (2.66), $\dot{\boldsymbol{r}}_2$ cannot be regarded as any exact differential. For this reason the task $\boldsymbol{r}_2$ is only defined by its time derivative.

Finally, let our *third manipulation variable* $\boldsymbol{r}_3 \in \Re^3$ be such as its time derivative is represented by the two planar velocities of the vehicle and its angular velocity around the vertical axis. They are identified as the components 4, 5 and 3 of the global quasivelocity vector:

$$\dot{\boldsymbol{r}}_3 = \boldsymbol{J}_3 \boldsymbol{p} = \begin{bmatrix} p_4 \\ p_5 \\ p_3 \end{bmatrix} \tag{4.23}$$

Hence:

$$\boldsymbol{J}_3 = \begin{bmatrix} 0\,0\,0\,1\,0\,0\,0\,0\,0\,0\,0\,0\,0 \\ 0\,0\,0\,0\,1\,0\,0\,0\,0\,0\,0\,0\,0 \\ 0\,0\,1\,0\,0\,0\,0\,0\,0\,0\,0\,0\,0 \end{bmatrix} \tag{4.24}$$

Even in this case, $\dot{\boldsymbol{r}}_3$ cannot be regarded as any exact differential (at least its third component), and the task $\boldsymbol{r}_3$ is only defined by its time derivative.

Note that, while closing the kinematic loop, the hovering condition is achieved by constantly setting the derivative of the third manipulation variable to zero:

$$\dot{\boldsymbol{r}}_3 = \boldsymbol{0}. \tag{4.25}$$

4.2.1.2 Experimental Data from the SAUVIM Recovery Task

The procedure described in Sect. 4.2.1.1 has been implemented and tested with the actual SAUVIM vehicle during a simulated recovery task. The experiment was carried out in the ocean at Snug Harbor, Honolulu, Hawaii (Fig. 4.4). The target shown in Fig. 4.5 was placed at a fixed underwater location. The experiment consisted in the following phases:

1. *Navigate the vehicle in proximity of the target.*
   The initial navigation was preformed manually, with the only goal of moving the vehicle to the target area.
2. *Place the vehicle in hovering just in front of the target.*
   The hovering configuration was chosen manually with the target at approximatively 2 m from the SAUVIM nose. The goal was to have the dipole at the border of

Fig. 4.4 SAUVIM navigating the water surrounding the area of the University of Hawaii Marine Center at Snug Harbor, Pier 45, Honolulu, Hawaii

Fig. 4.5 The visual-servoing target

the arm workspace, to allow the arm to reach a low manipulability configuration while visual-servoing to it.

3. *Extract the arm.*
   The arm was carrying the optical camera, capable of 6 DOF target pose detection.
4. *Start sweeping in search of the target.*
5. *Engage visual servoing.*
   After a successful lock, the visual servoing controller of the arm was set to main-

tain a fixed position and orientation of the end-effector w.r.t. the target, without performing any actual work. The visual servoing controller was the full 13 DOF kinematic controller described in Sect. 4.2.1.1.

In phase 5, the position of the target at the border of the workspace forced the controller to stretch the arm toward a low manipulability configuration, as shown in Fig. 4.6b.

Figure 4.7 show the measure of manipulability during the initial 50 s. In Fig. 4.7a, the augmented Jacobian is defined as:

$$\boldsymbol{J}_a = \begin{bmatrix} \boldsymbol{J}_1 \\ \boldsymbol{J}_2 \\ \boldsymbol{J}_3 \end{bmatrix} \tag{4.26}$$

with $\boldsymbol{J}_1$, $\boldsymbol{J}_2$ and $\boldsymbol{J}_3$ defined as in Eqs. (4.20), (4.22) and (4.24) respectively. It is noticeable at $t = 22$ s the acquisition of a stable lock of the target and the engagement of the visual servoing. As the arm reaches the target, the measure of manipulability decreases until its minimum for $t = 28$ s. From there, the task reconstruction tries to increase the manipulability by using the low 8 DOF (thrusters), which results in bringing the vehicle closer to the target, hence restoring the manipulability to acceptable values. This is also evidenced in Fig. 4.6c (a snapshot taken at $t = 50$ s), where the arm is in a better (less 'stretched') configuration compared to $t = 28$ s (Fig. 4.6b).

4.3 The SAUVIM Dynamic Control System

In the formulation of the workspace optimization we implicitly assumed that the SAUVIM control system provides a velocity input for each of its 8 joints schematized in Fig. 4.1. As shown in Sect. 3.1, the robotic arm electronic hardware already implements a closed loop velocity control at the corresponding joint (joints 2 through 8 of our UVMS model).

Instead, the necessary velocity input of the 6 DOF joint 1 is implemented within a dedicated dynamic controller, described in the following sections.

4.3.1 Vehicle Dynamics

In our study, the vehicle dynamics was modeled as described in Sect. 2.3. Here we are considering only the dynamics of the vehicle, assuming that the manipulator, during its operations, acts as a disturbance to the vehicle system. This is a reasonable assumption since the manipulator mass is only $1/100$ of the vehicle mass. With this assumption, the Lagrange equation for the quasi-coordinates introduced in

(a)

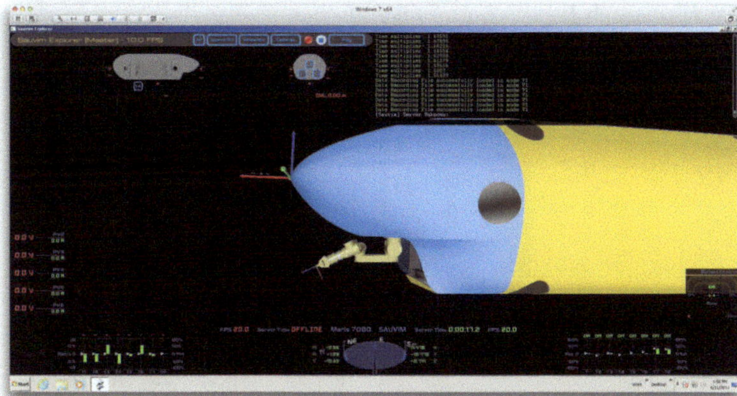

(b)

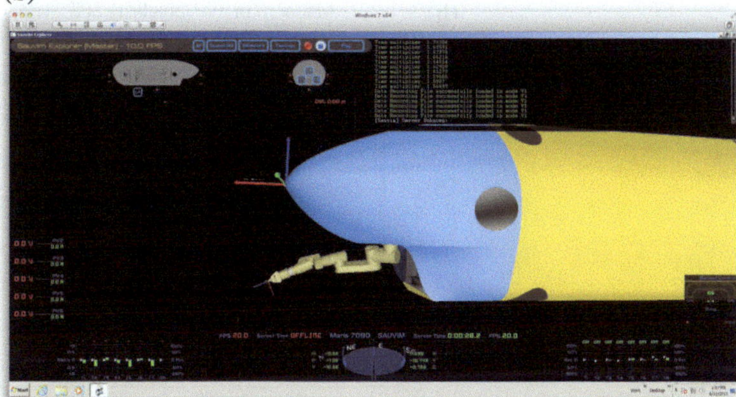

(c)

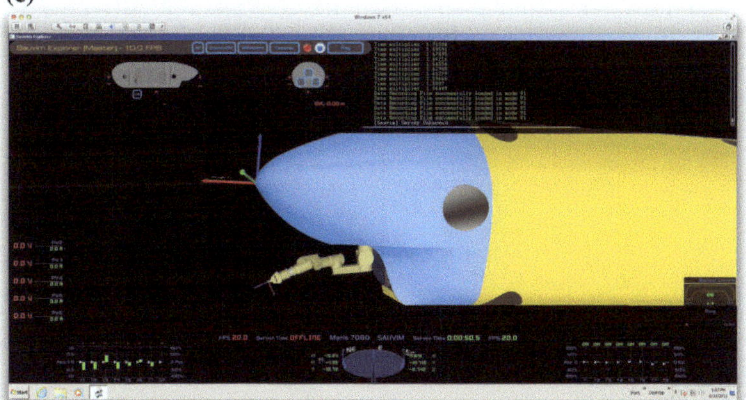

Fig. 4.6 SAUVIM configurations from experimental data results at **a** $t = 17\,\text{s}$, **b** $t = 28\,\text{s}$ and **c** $t = 50\,\text{s}$

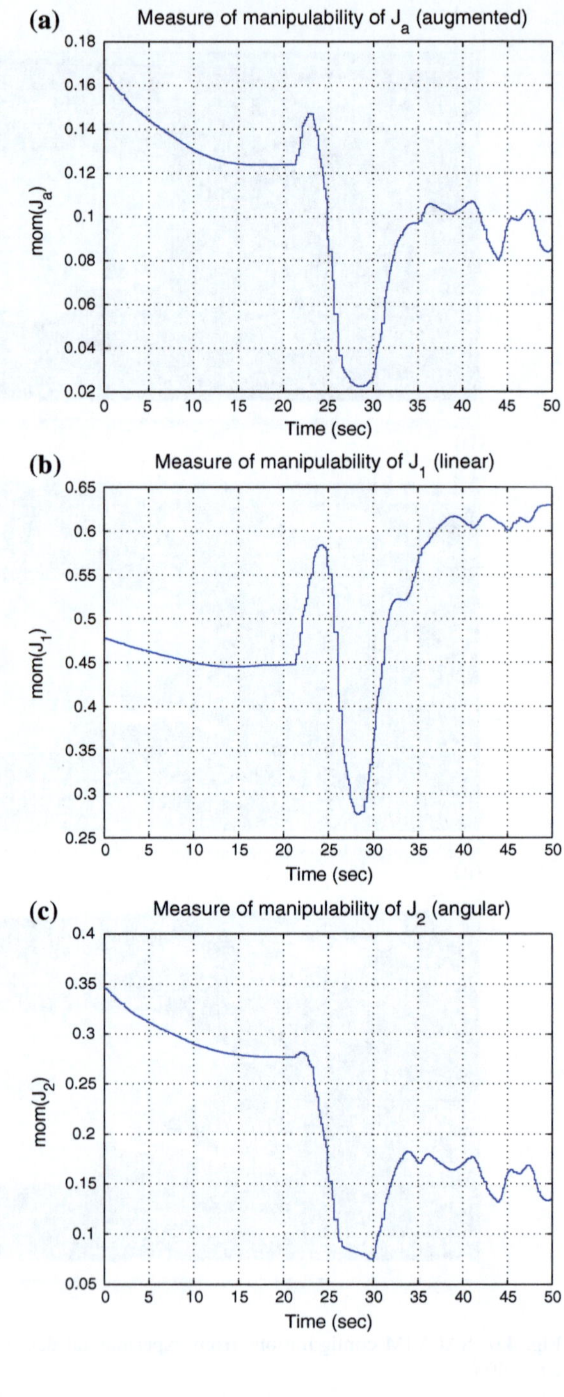

Fig. 4.7 Measure of manipulability of **a** $J_{augmented}$, **b** J_1 (*linear*) and **c** J_2 (*angular*), from experimental data

Eq. (2.190) becomes:

$$\boldsymbol{A}(\boldsymbol{q}_s)\dot{\boldsymbol{p}}_s + \boldsymbol{B}(\boldsymbol{q}_s, \boldsymbol{p}_s)\boldsymbol{p}_s = \boldsymbol{\mu}_e \tag{4.27}$$

where:

$\boldsymbol{q}_s$ is the vehicle configuration vector:

$$\boldsymbol{q}_s = \left[q_1, \ldots, q_6\right]^T = \left[\, roll\ pitch\ yaw\ x_{cm}\ y_{cm}\ z_{cm}\,\right]^T$$

with q_1 through q_6 defined in Eq. (4.1);

$\boldsymbol{p}$ is the system quasivelocity vector;

$$\boldsymbol{p}_s = \left[p_1, \ldots, p_6\right]^T = \left[\,{}^{cm}\omega_x\ {}^{cm}\omega_y\ {}^{cm}\omega_z\ {}^{cm}\boldsymbol{v}_x\ {}^{cm}\boldsymbol{v}_y\ {}^{cm}\boldsymbol{v}_z\,\right]^T$$

with p_1 through p_6 defined in Eq. (4.2);

$\boldsymbol{A}(\boldsymbol{q}_s)$ is the inertia matrix of the structure defined as in Eq. (2.180), and comprehensive of the added mass and added inertia;

$\boldsymbol{B}(\boldsymbol{q}_s, \boldsymbol{p}_s)$ is the is the matrix of Coriolis and centrifugal forces defined in Eq. (2.191) (also comprehensive of the added mass and added inertia);

$\boldsymbol{\mu}_e$ represents the projection in the space of the joint velocities (i.e. body axis in our case) of the external generalized forces.

The external generalized force vector $\boldsymbol{\mu}_e$ can be computed by recalling Eq. (2.169). Let $\boldsymbol{J}_s(\boldsymbol{q}_s, \boldsymbol{x})$ be the Jacobian of SAUVIM, such as:

$$\dot{\boldsymbol{X}}_P = \boldsymbol{J}_s(\boldsymbol{q}_s, \boldsymbol{x})\boldsymbol{p}_s \tag{4.28}$$

where $\dot{\boldsymbol{X}}_P$ is the generalized velocity (in the main frame) of the point P of coordinates $\boldsymbol{x}$ (expressed in the center of mass frame $\langle cm \rangle$). With this assumption, the projection in the space of the joint velocities of a generalized extern action $\boldsymbol{W}_P = \left[\boldsymbol{M}_{Px}\ \boldsymbol{M}_{Py}\ \boldsymbol{M}_{Pz}\ \boldsymbol{F}_{Px}\ \boldsymbol{F}_{Py}\ \boldsymbol{F}_{Pz}\right]^T$ applied on P of coordinates $\boldsymbol{x}$ is given by:

$$\boldsymbol{\mu}_P = \boldsymbol{J}_x^T(\boldsymbol{q}_s, \boldsymbol{x})\,\boldsymbol{W}_P \tag{4.29}$$

With this formalism, the generalized restoring force acting on the vehicle is given by:

$$\boldsymbol{\mu}_r = \boldsymbol{J}_s^T\left(\boldsymbol{q}_s, {}^{cm}\boldsymbol{x}_{cb}\right)\begin{bmatrix} 0 \\ 0 \\ 0 \\ 0 \\ 0 \\ \boldsymbol{F}_b \end{bmatrix} + \boldsymbol{J}_s^T(\boldsymbol{q}_s, \boldsymbol{0})\begin{bmatrix} 0 \\ 0 \\ 0 \\ 0 \\ 0 \\ \boldsymbol{g}\cdot m_S \end{bmatrix} \tag{4.30}$$

where ${}^{cm}\boldsymbol{x}_{cb}$ are the coordinates of the center of buoyancy projected in the center of mass frame $\langle cm \rangle$, $\boldsymbol{g}$ is the gravity acceleration, $\boldsymbol{F}_b$ is the resultant of the buoyancy force (applied to the COB) and m_S is the mass of SAUVIM.

In modeling our vehicle, since we consider the vehicle stationary in a hovering configuration, the external action is composed by the restoring force, linear damping actions and thruster forces. This leads to the final form of Eq. (4.27):

$$\boldsymbol{A}(\boldsymbol{q}_s)\dot{\boldsymbol{p}}_s + \boldsymbol{B}(\boldsymbol{q}_s, \boldsymbol{p}_s)\boldsymbol{p}_s = \boldsymbol{D} \cdot \boldsymbol{p}_s + TCM \cdot \boldsymbol{\tau}_T + \boldsymbol{\mu}_r \tag{4.31}$$

where $\boldsymbol{D}$ is a diagonal matrix associated with the linear damping term of the drag force, $\boldsymbol{\tau}_T \in \Re^8$ is the thrust vector and $TCM \in \Re^{6\times 8}$ is the thruster control matrix:

$$TCM = \begin{bmatrix} 0.7489 - cm_y & 1.537 + cm_x & 0 & 0\ 0\ 1 \\ -0.7489 - cmy & 1.537 + cm_x & 0 & 0\ 0\ 1 \\ 0.7489 - cm_y & 4.458 + cm_x & 0 & 0\ 0\ 1 \\ -0.7489 - cm_y & 4.458 + cm_x & 0 & 0\ 0\ 1 \\ 0 & -0.4826 - cm_z & -1.311 + cm_y & 1\ 0\ 0 \\ 0 & -0.4826 - cm_z & 1.311 + cm_y & 1\ 0\ 0 \\ 0.2413 + cm_z & 0 & -1.908 - cm_x & 0\ 1\ 0 \\ 0.2413 + cm_z & 0 & -5.153 - cm_x & 0\ 1\ 0 \end{bmatrix}^T \tag{4.32}$$

In the above matrix cm_x, cm_y and cm_z are the coordinates of the center of mass in the SAUVIM frame $\langle S \rangle$, parallel to the center of mass frame $\langle Cm \rangle$ and translated to the tip of the forward nose.

4.3.1.1 SAUVIM Dynamic Control

During a generic autonomous manipulation task the vehicle controller has the main responsibility of maintaining the vehicle in the desired configuration. For example, often the vehicle must be actively stabilized in a hovering configuration, while the manipulator performs its task. Among the hydrodynamic effects acting on a rigid body moving in a fluid, the restoring generalized forces (gravity plus buoyancy) and the ocean current are of major concern in designing a motion controller for intervention AUVs. In literature, several works [1–6] have been presented assessing the problem of 6 DOF control of AUVs. The effectiveness of the above works in compensating for the persistent dynamic effects, e.g., the restoring forces and the ocean current, was analyzed in [7].

The full 6 DOF SAUVIM control system was implemented using the control loop scheme introduced in Sect. 3.1.1, and illustrated in Fig. 3.2. In this case, the block named *Robot HW Controller* represents the dynamic control system of SAUVIM with generalized velocity input. It was implemented by using the standard computed torque controller.

Let's first define the auxiliary control input as:

$$\boldsymbol{a}_{ci} = \dot{\boldsymbol{p}}_{s-des} + \boldsymbol{K}_v\,(\boldsymbol{p}_{s-des} - \boldsymbol{p}_s) + \boldsymbol{K}_p\,(\boldsymbol{q}_{s-des} - \boldsymbol{q}_s) \tag{4.33}$$

where:

$\dot{\boldsymbol{p}}_{s-des}$ is the desired derivative of the quasivelocity vector (angular and linear acceleration s)
$\boldsymbol{p}_{s-des}$ is the desired quasivelocity vector
$\boldsymbol{q}_{s-des}$ is the desired configuration vector
$\boldsymbol{K}_v$ is the derivative matrix gain
$\boldsymbol{K}_p$ is the proportional matrix gain.

The computed torque, from Eq. (4.31), becomes:

$$\boldsymbol{\tau}_c = TCM \cdot \boldsymbol{\tau}_T = \boldsymbol{A} \cdot \boldsymbol{a}_{ci} + \boldsymbol{B} \cdot \boldsymbol{p}_s - \boldsymbol{D} \cdot \boldsymbol{p}_s - \boldsymbol{\mu}_r \tag{4.34}$$

By substituting Eq. (4.34) into Eq. (4.31) we can compute the dynamics of the error:

$$\dot{\boldsymbol{p}}_s - \dot{\boldsymbol{p}}_{s-des} + \boldsymbol{K}_v\,(\boldsymbol{p}_{s-des} - \boldsymbol{p}_s) + \boldsymbol{K}_p\,(\boldsymbol{q}_{s-des} - \boldsymbol{q}_s) = \boldsymbol{0} \tag{4.35}$$

Note that, in general, in all the case where $\boldsymbol{p} = \dot{\boldsymbol{q}}$ (for example in case of 1 DOF rotational joints), Eq. (4.35) describes a second order stable dynamics, with real eigenvalues (the eigenvalues of $\boldsymbol{K}_p$ and $\boldsymbol{K}_v$).

The SAUVIM kinematic controller is then derived by Eqs. (4.33) and (4.34) by simply putting $\boldsymbol{q}_{s-des} - \boldsymbol{q}_s = \boldsymbol{0}$ and computing $\dot{\boldsymbol{p}}_{s-des}$ by numerical derivation of $\boldsymbol{p}_{s-des}$. This operation makes Eq. (4.35) stable even if $\boldsymbol{p} \neq \dot{\boldsymbol{q}}$, which is the case being here considered.

4.4 Identification of Dynamic Parameters

As it appears evident within its formulation, the computed torque controller needs the knowledge of the system dynamic parameters. In particular we need:

- Mass
- Center of Mass location
- Inertia tensor
- Center of buoyancy
- Added mass.

In general, one of the main problems is the lack of knowledge of the restoring-related dynamic parameters, especially in case of heavy vehicles. The location of the center of buoyancy (COB) with respect to the center of mass (COM) plays a fundamental role in the performance of the dynamic control. The importance of its knowledge is also related to the problem of power optimization, since very often a working orientation of the vehicle does not have strict constrains, being the manipulator more capable of realizing specific orientations than the vehicle. Thus a working vehicle orientation could be simply chosen as the one which aligns the COB above the COM, minimizing the power requirement for maintaining the vehicle in hovering.

In SAUVIM we have developed a methodology for identifying the relative position of the COB with respect to the center of mass. This approach is based on the use of an Extended Kalman Filter (EKF), and being suitable of real-time implementation, it responds to any change in the vehicle configuration (i.e. ballast operations or manipulator dynamics).

This algorithm was successfully tested first in a simulation environment and then on the actual vehicle with the model-based dynamic control presented in Sect. 4.3.1.1.

4.4.1 COB Identification with Extended Kalman Filter

The state space model of our Extended Kalman Filter must contains information on the vehicle dynamics. Equation (4.31) describes the evolution of the generalized position, velocity and acceleration of the vehicle through time. It can be solved with respect to the derivative of the quasi-velocity vector as:

$$\dot{\boldsymbol{p}} = \boldsymbol{A}(\boldsymbol{q})^{-1}\left(-\boldsymbol{B}(\boldsymbol{q},\boldsymbol{p})\,\boldsymbol{p} + \boldsymbol{D}\cdot\boldsymbol{p} + TCM\cdot\boldsymbol{\tau} + \boldsymbol{\mu}_r\right) \tag{4.36}$$

The space equations of the EKF are then augmented with 4 more equations, describing the evolution of the three coordinates of the center of buoyancy and the buoyancy force. Since the ballast movements are much slower than the dynamics of the vehicle, their evolution is simply given by (see Eq. 4.30):

$$\begin{cases} {}^{cm}\dot{\boldsymbol{x}}_{cb} = \boldsymbol{0} \\ \dot{\boldsymbol{F}}_b = \boldsymbol{0} \end{cases} \tag{4.37}$$

We also need to add the evolution of the generalized position of the vehicle. This is easily done by integrating the derivative of the rotation matrix ${}^{0}_{cm}\dot{\boldsymbol{R}}$ of the center of mass frame $\langle cm\rangle$ w.r.t. the main frame $\langle 0\rangle$:

$${}^{0}_{cm}\dot{\boldsymbol{R}} = {}^{0}_{cm}\boldsymbol{R}\begin{bmatrix} 0 & -p_3 & p_2 \\ p_3 & 0 & -p_1 \\ -p_2 & p_1 & 0 \end{bmatrix} \tag{4.38}$$

and the linear velocity of the center of mass ${}^{0}\dot{\boldsymbol{x}}_{cm} = \begin{bmatrix} \dot{x}_{cm} & \dot{y}_{cm} & \dot{z}_{cm} \end{bmatrix}^T$ projected in the main frame $\langle 0\rangle$:

$$ {}^{0}\dot{\boldsymbol{x}}_{cm} = {}^{0}_{cm}\boldsymbol{R} \begin{bmatrix} p_4 \\ p_5 \\ p_6 \end{bmatrix} \tag{4.39} $$

where we assumed that $\boldsymbol{p}_s = \begin{bmatrix} p_1 \; p_2 \; p_3 \; p_4 \; p_5 \; p_6 \end{bmatrix}^T$.

4.4.1.1 Process Equations

The complete non-linear process can be described by joining together Eqs. (4.36), (4.37), (4.38) and (4.39), and adding the appropriate noise.

In the formulation of the vehicle dynamic model (4.31) we have implicitly assumed that the center of mass is fixed w.r.t. the vehicle body. This may not be necessarily true, especially during ballast operations, and its variation (despite small) may affect the computation of the inertia tensor and the thruster control matrix TCM. This issue, together with the fact that several dynamic parameters are known approximatively, is considered in our formulation by introducing appropriate noise terms w_i to the following variables:

- Total mass:

$$ \bar{m}_t = m_t + w_{mt} \tag{4.40} $$

- Total inertia tensor:

$$ \bar{\boldsymbol{I}}_t = \boldsymbol{I}_t + \begin{bmatrix} w_{i11} & w_{i12} & w_{i13} \\ w_{i12} & w_{i22} & w_{i23} \\ w_{i13} & w_{i23} & w_{i33} \end{bmatrix} \tag{4.41} $$

- Location of the center of mass ${}^{S}\boldsymbol{x}_{cm}$ projected in the main SAUVIM frame $\langle S \rangle$ (for the computation of the TCM):

$$ {}^{S}\bar{\boldsymbol{x}}_{cm} = {}^{S}\boldsymbol{x}_{cm} + \begin{bmatrix} w_{cmx}, \, w_{cmy}, \, w_{cmz} \end{bmatrix}^T \tag{4.42} $$

- Thrust vector (considering that our vehicle has 8 thrusters):

$$ \bar{\boldsymbol{\tau}} = \boldsymbol{\tau} + [w_{t1}, \, \ldots, \, w_{t8}]^T \tag{4.43} $$

- Damping coefficients (with k_d an estimated constant):

$$ \bar{\boldsymbol{D}} = diag\,(k_d + w_{kd}) \tag{4.44} $$

With these assumptions, after discretizing the system Eqs. (4.36) and (4.37), the non-linear stochastic difference equation of the extended Kalman filter becomes:

$$\boldsymbol{x}_k = \begin{bmatrix} \boldsymbol{p}_k \\ ({}^{cm}\boldsymbol{x}_{cb})_k \\ (\boldsymbol{F}_b)_k \\ \left({}^{0}_{cm}\boldsymbol{R}\right)_k \\ \left({}^{0}\boldsymbol{x}_{cm}\right)_k \end{bmatrix}$$

$$= \begin{bmatrix} \boldsymbol{p}_{k-1} + \bar{\boldsymbol{A}}^{-1}\left(-\bar{\boldsymbol{B}}\boldsymbol{p}_{k-1} + \bar{\boldsymbol{D}}\boldsymbol{p}_{k-1} + T\bar{C}M\bar{\boldsymbol{\tau}} + \boldsymbol{\mu}_r\right)\Delta T \\ ({}^{cm}\boldsymbol{x}_{cb})_{k-1} + (\boldsymbol{w}_{cb})_{k-1} \\ (\boldsymbol{F}_b)_{k-1} + (\boldsymbol{w}_{Fb})_{k-1} \\ \left({}^{0}_{cm}\boldsymbol{R}\right)_{k-1} + \left({}^{0}_{cm}\boldsymbol{R}\right)_{k-1} \begin{bmatrix} 0 & -p_3 & p_2 \\ p_3 & 0 & -p_1 \\ -p_2 & p_1 & 0 \end{bmatrix}_{k-1} \Delta T \\ \left({}^{0}\boldsymbol{x}_{cm}\right)_{k-1} + \left({}^{0}_{cm}\boldsymbol{R}\right)_{k-1} \left(\left[p_4\ p_5\ p_6\right]^T\right)_{k-1} \Delta T \end{bmatrix} \tag{4.45}$$

where ΔT is the sample time. Note that many noise variables are encapsulated within $\bar{\boldsymbol{A}}$, $\bar{\boldsymbol{B}}$, $T\bar{C}M$ and $\bar{\boldsymbol{\tau}}$. The last, vector of the thrust forces, is the process input. The dimension on the state vector is thus 22.

4.4.1.2 Measurement Equations

As mentioned in Sect. 1.2, the SAUVIM navigation sensor system includes a Photonic Inertial Navigation System (PHINS) unit from IXSEA (Fig. 4.8). This INS unit produces a full set of position, orientation, linear and angular velocities, linear and angular accelerations. Its high accuracy of inertial measurement capability is based on Fiber Optic Gyroscope technology and an embedded digital signal processor that runs an Extended Kalman Filter (EKF). In order to improve the accuracy of the vehicle position, further measurement inputs are provided to the EKF of the PHINS by a set of external sensors: a differential GPS, a Doppler Velocity Log (DVL) and the classical depth sensor. The outputs of the PHINS are the following:

- ${}^{0}_{Ph}\boldsymbol{T}$ generalized position of the PHINS w.r.t. the main frame $\langle 0 \rangle$;
- ${}^{0}\boldsymbol{v}_{Ph}$ linear velocity of the PHINS projected on the main frame $\langle 0 \rangle$;
- ${}^{Ph}\boldsymbol{\omega}_{Ph}$ angular velocity of the PHINS projected on the PHINS frame $\langle Ph \rangle$;
- ${}^{Ph}\boldsymbol{a}_{Ph}$ linear acceleration the PHINS projected on the PHINS frame $\langle Ph \rangle$;

The above outputs represent the measurement inputs of our SAUVIM EKF. The measurement equations are more complex than the process evolution: this is due to the fact that the PHINS is physically mounted with an offset, transitional and rotational, w.r.t. the center of mass (see Fig. 4.9).

In order to integrate them in the EKF it is necessary to express the quasi-velocity vector and its derivative in term of the PHINS outputs.

The *transformation matrix* ${}^{0}_{Ph}\boldsymbol{T}$ of the PHINS frame w.r.t. the main frame can be easily expressed as follows:

$${}^{0}_{Ph}\boldsymbol{T} = {}^{0}_{cm}\boldsymbol{T} \cdot {}^{cm}_{Ph}\boldsymbol{T} \tag{4.46}$$

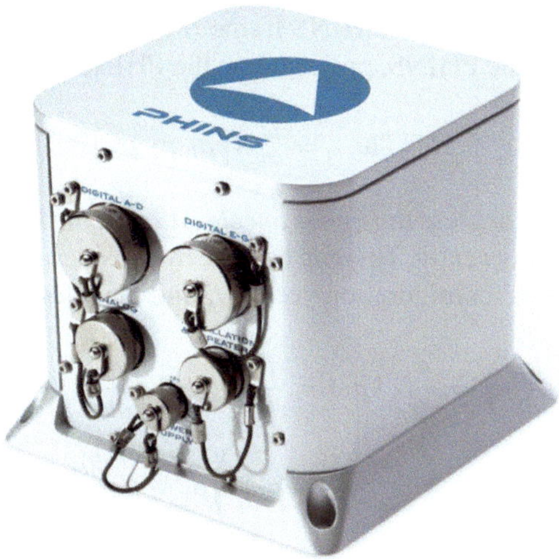

Fig. 4.8 The Photonic Inertial Navigation System PHINS (courtesy by IXSEA)

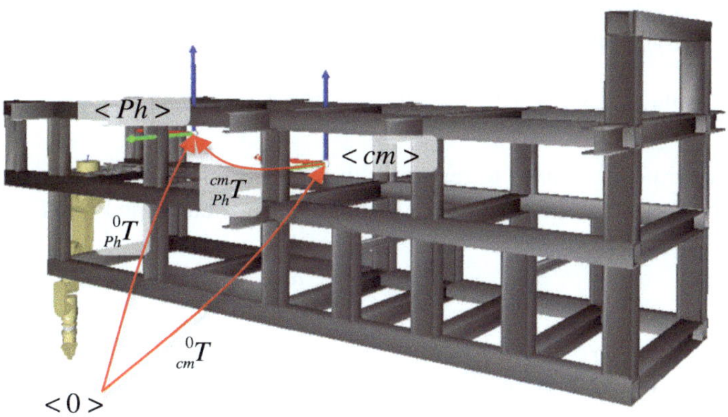

Fig. 4.9 Placement of the INS in SAUVIM

which is a function of the transformation matrix of the center of mass, the state variables ${}^{0}_{cm}\boldsymbol{R}$, ${}^{0}\boldsymbol{x}_{cm}$ and of the known generalized offset ${}^{cm}_{Ph}\boldsymbol{T}$ (see Fig. 4.9). The reorganization of the transformation matrix ${}^{0}_{Ph}\boldsymbol{T}$ into a 12 element vector represents the first set of measurement equations.

The *linear velocity* of the PHINS, projected on the main frame $\langle 0 \rangle$, can be computed from the quasi-velocity vectors making use of the linear part of the Jacobian (i.e. the bottom-half of the matrix) of Eq. (4.28):

$$ {}^{0}\boldsymbol{v}_{Ph} = {}^{0}\boldsymbol{J}_{Lin}\,(\boldsymbol{q}_s, \boldsymbol{r}_{Ph})\,\boldsymbol{p}_s \tag{4.47} $$

where $\boldsymbol{r}_{Ph}$ is the location of the origin PHINS frame $\langle Ph \rangle$ w.r.t. the main frame $\langle 0 \rangle$.

The *angular velocity* of the PHINS, projected on the PHINS frame $\langle P \rangle$, is simply given by:

$$^{Ph}\boldsymbol{\omega}_{\mathbf{Ph}} = {}^{Ph}_{cm}\boldsymbol{R} \cdot {}^{cm}\boldsymbol{\omega}_{cm} = {}^{Ph}_{cm}\boldsymbol{R}\left[p_1\ p_2\ p_3\right]^T \tag{4.48}$$

Finally, it is necessary to express the *linear acceleration* of the PHINS in terms of derivative of the quasivelocity vector. This can be done by considering the relationship between the derivative of the configuration vector $\boldsymbol{q}_s$ and the quasi-velocity $\boldsymbol{p}_s$:

$$\begin{aligned}\frac{d}{dt}\left(\frac{d\boldsymbol{q}_s}{dt}\right) &= \frac{d}{dt}\left(\boldsymbol{V}\boldsymbol{p}_s\right) = \frac{d\boldsymbol{V}}{dt}\boldsymbol{p}_s + \boldsymbol{V}\frac{d\boldsymbol{p}_s}{dt} \\ &= \left(\sum_{i=1}^{6}\frac{\partial \boldsymbol{V}}{\partial q_i}\left[\boldsymbol{V}\boldsymbol{p}_s\right]_i\right)\boldsymbol{p}_s + \boldsymbol{V}\frac{d\boldsymbol{p}_s}{dt}\end{aligned} \tag{4.49}$$

where $\boldsymbol{V}(\boldsymbol{q}_s)$ is the transformation matrix between the $\boldsymbol{p}_s$ space and the $\dot{\boldsymbol{q}}_s$ space, such as $\dot{\boldsymbol{q}}_s = \boldsymbol{V}\boldsymbol{p}_s$ (see Eq. 2.192).

The last three elements of the vector (4.49) represent the linear acceleration $^0\boldsymbol{a}_{cm}$ of the center of mass projected on the main frame $\langle 0 \rangle$. This is related to the linear acceleration $^0\boldsymbol{a}_{Ph}$ of the PHINS frame, projected on the main frame, by the following relation:

$$^0\boldsymbol{a}_{Ph} = {}^0\boldsymbol{a}_{cm} + {}^0\boldsymbol{v}_{cm} \times {}^0\boldsymbol{r}_{Ph} + {}^0_{cm}\boldsymbol{R}\left[\dot{p}_1\ \dot{p}_2\ \dot{p}_3\right]^T \times {}^0\boldsymbol{r}_{Ph} \tag{4.50}$$

where the operator $\times$ is used to represent the cross product between two vectors.

The linear acceleration (4.50) must be finally projected to the PHINS frame $\langle Ph \rangle$:

$$^{Ph}\boldsymbol{a}_{Ph} = {}^{Ph}_{0}\boldsymbol{R}\,{}^0\boldsymbol{a}_{Ph} \tag{4.51}$$

Note that the quantity $\frac{d\boldsymbol{p}_s}{dt}$ can be easily computed from the first 6 state Eq. (4.36).

Finally, combining together Eqs. (4.46), (4.47), (4.48) and (4.50) we obtain a set of 21 equations which represent the measurement equations of the extended Kalman filter:

$$z_k = \begin{bmatrix} \left({}^{Ph}_{0}\boldsymbol{T}_{1..3}\right)_k + \boldsymbol{\nu}_{Tk} \\ \left({}^0\boldsymbol{v}_{Ph}\right)_k + \boldsymbol{\nu}_{vk} \\ \left({}^{Ph}\boldsymbol{\omega}_{Ph}\right)_k + \boldsymbol{\nu}_{\omega k} \\ \left({}^{Ph}\boldsymbol{a}_{Ph}\right)_k + \boldsymbol{\nu}_{ak} \end{bmatrix} \tag{4.52}$$

The vectors $\boldsymbol{\nu}_{Tk}$, $\boldsymbol{\nu}_{vk}$, $\boldsymbol{\nu}_{\omega k}$ and $\boldsymbol{\nu}_{ak}$ represent the measure noise, associated with the PHINS unit and comprehensive of the uncertainty of the location of the center of mass w.r.t. the PHINS unit (the generalized offset ${}^{cm}_{Ph}\boldsymbol{T}$ in Eq. 4.46).

4.4.2 Optimal Configuration for Hovering

The importance of the knowledge of the center of buoyancy is also related to the problem of power optimization. During an autonomous manipulation task, the working orientation of the vehicle does not have strict constraints, being the manipulator more capable of realizing specific orientations than the vehicle. Thus a working vehicle orientation could be simply chosen as the one that aligns the COB above the COM, minimizing the power requirement for maintaining the vehicle in hovering. The equilibrium target rotation is thus computed with the following considerations.

Let the rotation matrix of the equilibrium configuration be:

$$ {}^{0}_{e}\boldsymbol{R} = \left[{}^{0}\boldsymbol{i}_e \; {}^{0}\boldsymbol{j}_e \; {}^{0}\boldsymbol{k}_e \right] \tag{4.53}$$

To reach our goal, the axis ${}^{0}\boldsymbol{k}_e$ is chosen to be parallel to the COM-COB segment:

$$ {}^{0}\boldsymbol{k}_e = \frac{{}^{0}\boldsymbol{x}_{cb}}{\left|{}^{0}\boldsymbol{x}_{cb}\right|} \tag{4.54}$$

The axis ${}^{0}\boldsymbol{i}_e$ is instead chosen as the orthogonal to the plane formed by ${}^{0}\boldsymbol{k}_e$ and the axis $\left[0\;1\;0\right]^T$:

$$ {}^{0}\boldsymbol{i}_e = \frac{\left[0\;1\;0\right]^T \times {}^{0}\boldsymbol{k}_e}{\left|\left[0\;1\;0\right]^T \times {}^{0}\boldsymbol{k}_e\right|} \tag{4.55}$$

Finally, the ${}^{0}\boldsymbol{j}_e$ axis is simply computed as the cross product:

$$ {}^{0}\boldsymbol{j}_e = {}^{0}\boldsymbol{k}_e \times {}^{0}\boldsymbol{i}_e \tag{4.56}$$

The matrix (4.53) is then transformed in roll, pitch and yaw angles and the first two replace the target values of the navigation controller.

4.4.3 Implementation

The complex nature of the EKF described above is the main obstacle to its practical implementation. In SAUVIM, this was overcomed by first creating a set of automated tools for code generation. The computation of the partial derivatives of Eqs. (4.45) and (4.52), necessary for the implementation of the extended Kalman filter, was done using symbolic computation packages. The symbolic processor was then used for generating the 5,000+ lines of C++ code necessary to import the EKF in our hardware system.

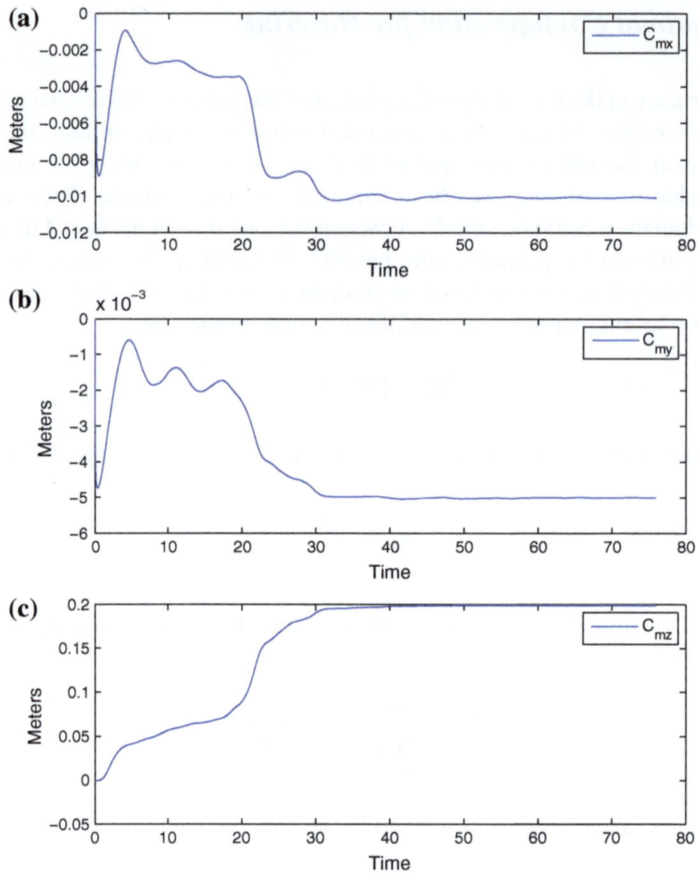

Fig. 4.10 Output of the EKF during random navigation over 70 s. **a** x-coordinate of the center of buoyancy, **b** y-coordinate of the center of buoyancy, **c** z-coordinate of the center of buoyancy

Despite the complexity and length, with the optimization introduced by the code generation process, the release version of the EKF runs at the sampling rate of about 50 ms, hence perfectly suitable to an online implementation in the vast majority of navigation controllers.

4.4.4 Simulation Results

The initial results of the EKF for COB identification have been validated using the SAUVIM simulator. The simulator implements the Lagrange equation for quasi-coordinates (4.31), inclusive of the drag, restoring force and added mass.

Fig. 4.11 SAUVIM during hovering

The first simulation consisted in performing simple open-loop oscillations around the three center of mass axis, in order to assess the performance of the EKF. Results are shown in Fig. 4.10. Here, the three coordinates of the center of buoyancy successfully converge toward the values of the model ($x = -10$ mm, $y = -5$ mm and $z = 200$ mm). The accuracy, in this case, is in the order of fraction of a millimeter.

4.4.5 Experimental Results

The feasibility of the EKF for COB identification was experimentally confirmed with the SAUVIM vehicle. In this case the vehicle is set to maintain a hovering position in shallow water, with the roll and pitch that were chosen such that the COB is aligned over the COM. Figure 4.11 is a snapshot taken during the experiment with a remotely operated underwater camera, showing the bottom frame of SAUVIM hovering above its docking platform (the metallic structure at the bottom of Fig. 4.11).

In this experiment the dynamic vehicle controller is the one presented in Sect. 4.3.

Figure 4.12a and b show the outputs of the EKF for the x and y coordinates. During this experiment, the COB location was moved at $t = 10$ s, by filling the front ballast and emptying the rear one of approximately the same mass of water, while the global buoyancy was set to neutral. The location of the COB is visibly shifted forward along x, while it is almost unchanged along y. As seen in Fig. 4.13, the forward drift of the center of buoyancy created a large restoring torque that the position controller tried to compensate for by increasing the thrust action. After the Kalman filter converged toward the new estimate of the COB, the global module of

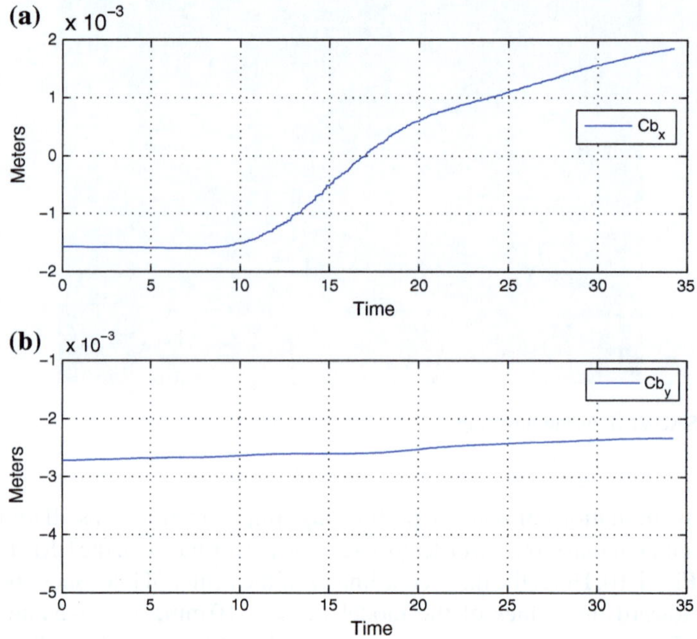

Fig. 4.12 Output of the EKF during hovering

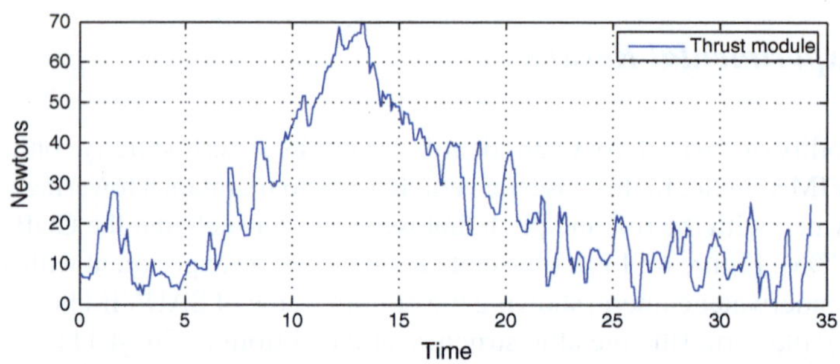

Fig. 4.13 Module of the thrust vector during hovering

the thrust vector successively converged toward zero, thus confirming that the newly computed working pitch angle was such that the COB was perfectly aligned over the COM. Figure 4.14 shows the change in roll and pitch introduced by the position controller and computed as explained in Sect. 4.4.2.

In order to compute the estimate of thruster force we used an approach based on simultaneous measurement of voltage, current and velocity, as described in [8].

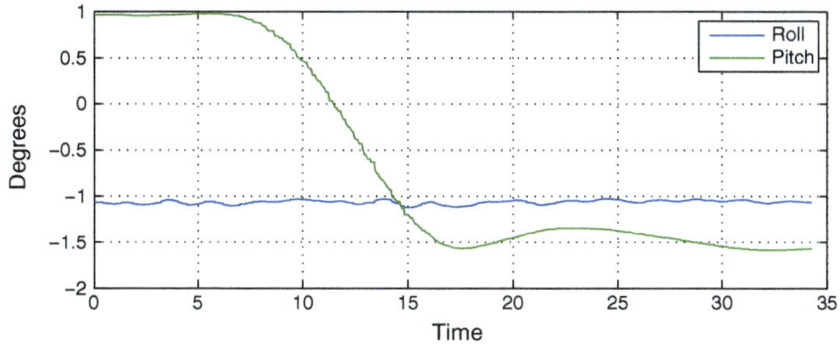

Fig. 4.14 Roll and pitch angles of SAUVIM during hovering

Note that, with this vehicle, even a small fraction of degree in the target orientation would generate a compensation thrust of several kilograms, given the mass of 3,660 kg and 90 mm of distance between the COB and COM of the vehicle. This is clearly confirmed by the previous plots, where a shift of about 3 mm of the x coordinate of the COB generated about 70 newtons of total force in the thrusters for compensating for the restoring force.

References

1. Antonelli G, Caccavale F, Chiaverini S, Fusco G (2001) A novel adaptive control law for autonomous underwater vehicles. In: Proceedings of IEEE international conference on robotics and automation (ICRA 2001), vol 1, pp 447–452. doi:10.1109/ROBOT.2001.932591
2. Fjellstad OE, Fossen T (1994) Position and attitude tracking of auv's: a quaternion feedback approach. IEEE J Oceanic Eng 19(4):512–518. doi:10.1109/48.338387
3. Fjellstad OE, Fossen T (1994) Singularity-free tracking of unmanned underwater vehicles in 6 dof. In: Proceedings of the 33rd IEEE conference on decision and control, vol 2, pp 1128–1133. doi:10.1109/CDC.1994.411068
4. Fossen T, Balchen J (1991) The nerov autonomous underwater vehicle. In: Proceedings of ocean technologies and opportunities in the Pacific for the 90's (OCEANS '91), pp 1414–1420
5. Sun Y, Cheah C (2003) Adaptive setpoint control for autonomous underwater vehicles. In: Proceedings of 42nd IEEE conference on decision and control, vol 2, pp 1262–1267. doi:10.1109/CDC.2003.1272782
6. Yuh J, Nie J, Lee C (1999) Experimental study on adaptive control of underwater robots. In: Proceedings of 1999 IEEE international conference on robotics and automation, vol 1, pp 393–398. doi:10.1109/ROBOT.1999.770010
7. Antonelli G (2006) Underwater robots, 2nd edn. Springer, Berlin
8. Hanai A, Choi S, Marani G, Rosa K (2009) Experimental validation of model-based thruster fault detection for underwater vehicles. In: IEEE international conference on robotics and automation (ICRA '09) , pp 194–199. doi:10.1109/ROBOT.2009.5152425

Chapter 5
Target Localization

This Chapter focuses on the methodology for locating the underwater target, driving the vehicle toward it and performing autonomous manipulation. Our approach investigates the feasibility of medium-range target identification using the DIDSON sonar, which was demonstrated during one of the first experiments that used the DIDSON for autonomous real-time target detection. Current results represent an important passage toward the development of a higher level of autonomy for intervention AUVs, providing advanced engineering solution to many new underwater tasks and applications that the fly-by type AUVs have not been able to handle.

5.1 Target Identification and Localization

One of the most difficult aspects of autonomous intervention missions is the identification and localization of the target. The localization subsystem for the autonomous manipulation of SAUVIM, employs different technologies (acoustical and optical) to guarantee a suitable, range dependent level of reliability, precision and accuracy. The SAUVIM system relies on three main sensing methods in order to acquire reliable data. As shown in Fig. 5.1, different sensor technologies were used, depending on the required range and accuracy.

For *long range* distance (over 30 m), SAUVIM utilizes the 375 kHz image sonars for searching an object and directing the vehicle toward the target zone.

For *mid-range* distance (2–40 m), the DIDSON sonar (Fig. 5.2) is used for a more accurate recognition and localization of the target. This is the phase where the vehicle has to position itself to have the target confined within the manipulation workspace.

Finally, when the target is within the manipulator workspace, *short range* and high accuracy sensors are used to perform the actual intervention task. This goal is achieved with the combined use of underwater video cameras and an ultrasonic motion tracker to retrieve the real-time six DOF position of the target during the manipulation tasks as reported in the literature [1] and [2].

G. Marani and J. Yuh, *Introduction to Autonomous Manipulation*,
Springer Tracts in Advanced Robotics 102, DOI: 10.1007/978-3-642-54613-6_5,

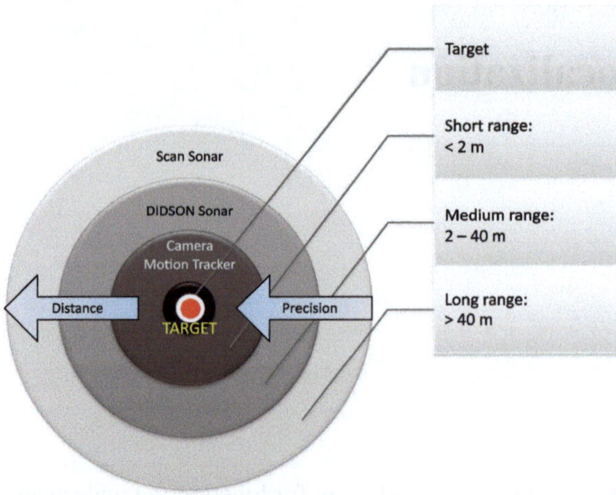

Fig. 5.1 The phases involved in a search for the target

Fig. 5.2 The DIDSON sonar (courtesy Sound Metrics Corp.)

A related work was presented in [3]. However, they assumed a structured environment and used self-similar landmarks (SSL, see [4]) as a robust, color, scale, and rotational invariant means of target identification from which the information could be used to guide an AUV for short-range homing and docking operations.

5.2 Mid-Range Object Identification with DIDSON Sonar

The DIDSON (Dual frequency IDentification SONar, Fig. 5.2) is a sonar with acoustic lenses, which operates at two frequencies, 1 and 1.8 MHz, with an operative range up to 40 m [5]. The medium range target localization using the DIDSON sonar is still a challenging task under development. Few works have been presented in literature on target identification and localization with the DIDSON. A preliminary study on target identification was presented in [6, 7], while a possible use of the DIDSON imagery for SLAM was investigated in [8].

The display method of the DIDSON sonar differs substantially from the optical camera. In general, a point in the space can be expressed in either Cartesian or spherical coordinates, (x, y, z) and (ρ, θ, φ) respectively. Indicating with $\Phi = \frac{\pi}{2} - \varphi$ the elevation, the two coordinate systems are related by (Fig. 5.3):

$$\begin{bmatrix} x \\ y \\ z \end{bmatrix} = \begin{bmatrix} \rho \cos(\Phi) \cos(\theta) \\ \rho \cos(\Phi) \sin(\theta) \\ \rho \sin(\Phi) \end{bmatrix} \tag{5.1}$$

The image in optical cameras follows projective geometry principles, and any point along the ray going through the optical center projects to the same image point. A generic 3D point (x, y, z) maps to the point (x_c, y_c) of the camera image and the information about the distance from the camera origin is lost.

The DIDSON, instead, records the acoustic refractance from an azimuth direction θ with a range measurement ρ. Hence, the generic 3D point in spherical coordinates (ρ, θ, φ) maps to the 2D point (ρ_d, θ_d). In this case, the information φ is lost. In every vertical plane passing through the center of the DIDSON, a point in the space at the same distance from the origin maps to the same point in the image plane.

In other terms, the relationship between the 3D Cartesian space and the 2D DIDSON image is given by solving Eq. (5.1):

$$\begin{bmatrix} \rho_d \\ \theta_d \end{bmatrix} = \begin{bmatrix} \sqrt{x^2 + y^2 + z^2} \\ \arctan(y, x) \end{bmatrix} \tag{5.2}$$

where (x, y, z) are the Cartesian coordinates of a generic point in 3D space and (ρ_d, θ_d) are the coordinate in the 2D DIDSON image. Another geometrical interpretation of Eq. (5.2) is given in [6]. In our experiment, the target was set as the platform in Fig. 5.4. The raw polar image acquired by the DIDSON is shown in Fig. 5.5a. Sometime it is preferable to display the image in terms of rectangular coordinates:

$$(x_s, y_s) = (\rho_d \sin \theta_d, \rho_d \cos \theta_d) \tag{5.3}$$

Figure 5.5b shows the same image of Fig. 5.5a after the transformation in rectangular coordinates.

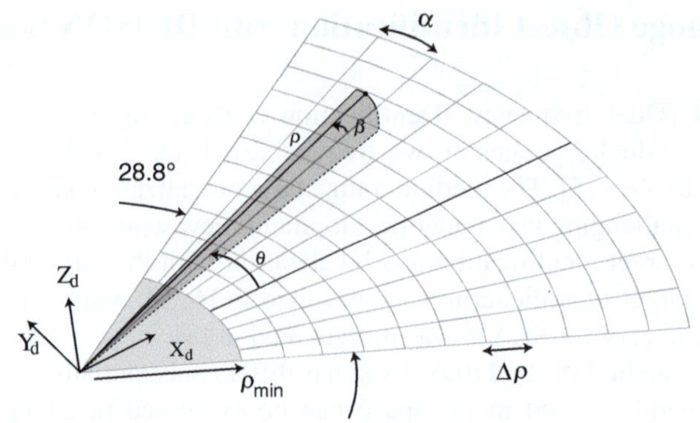

Fig. 5.3 Schematic representation of the DIDSON field of view

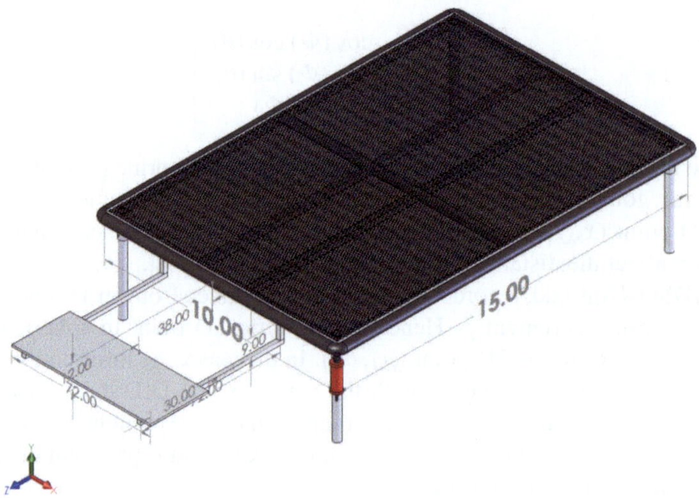

Fig. 5.4 The target in our search (dimensions are here given in feet)

Since the acquired image is extremely noisy, a feature extraction employing standard techniques like edge detection is rather problematic. In order to maximize the signal-noise ratio, we used a multi-dimensional matching filter to estimate the position of the target in the acquired image. The target identification and localization process is performed by iterating the steps described in the following sections.

5.2.1 Model Building

This phase involves the creation of the model of the target, as ideally seen from the DIDSON. Supposed that the generalized relative position (orientation and transla-

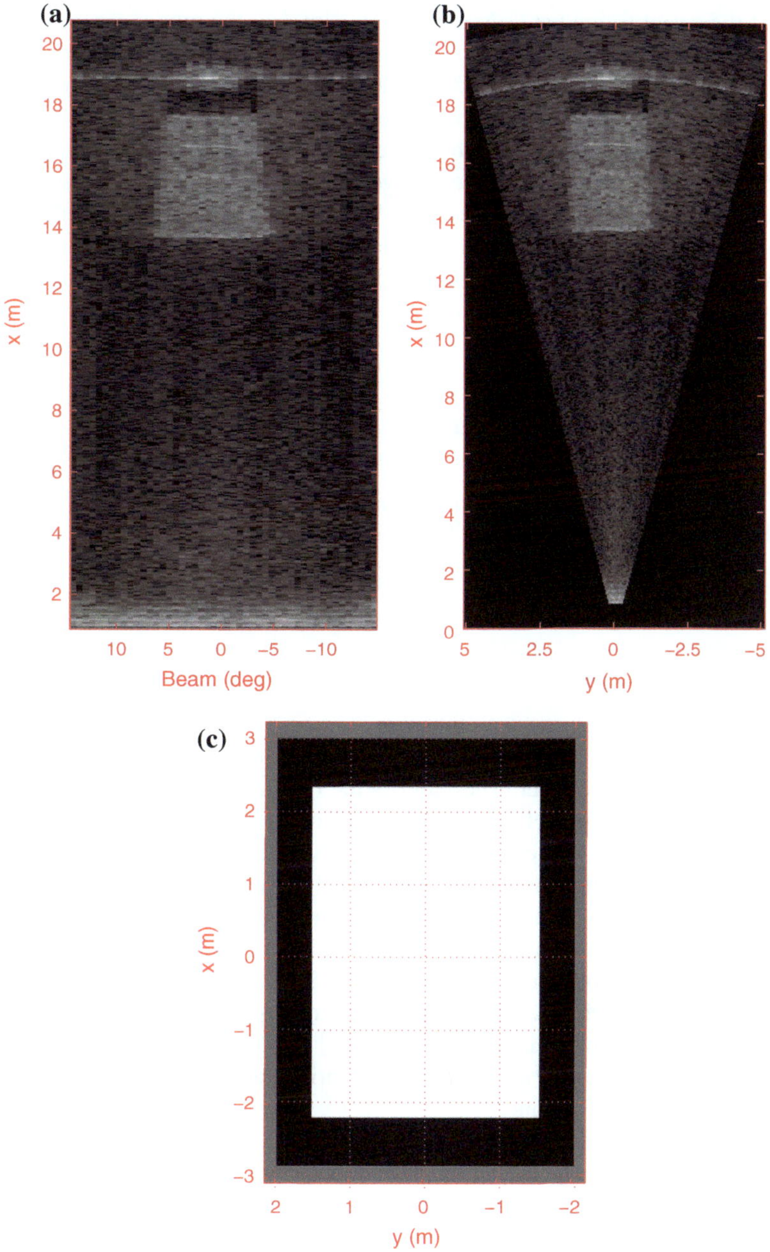

Fig. 5.5 Images from the DIDSON of the target platform, and relative computed model. **a** The target in raw polar coordinates, **b** The target in rectangular coordinates, **c** The generated model image in Cartesian space

tion) of the target frame w.r.t. the DIDSON is known, the model can be built using simple ray-tracing techniques, with maximum reflectivity (white) in correspondence of the surface and zero (grey) outside, with a shadow (black) area of few pixels added around the contour. Figure 5.5c shows an example model of the target, computed for the particular relative positions of the vehicle and platform for the observation shown in Fig. 5.5b. Also, since the shape of the generated model depends (even if slightly) on the position of the target with respect to the DIDSON, it is assumed that at the initial step the target is located at the center of its viewing area. As described below, the overall procedure of model building has to be refreshed in real-time with the updated vehicle positions.

5.2.2 Image Acquisition and Filtering

The DIDSON image is being captured and processed via OpenCV [9], in order to reduce some amount of noise. This is accomplished by a convolution with the Gaussian kernel.

5.2.3 Matched Filter

The model built from the previous step is searched within the processed DIDSON image using a matched filter. The latter is the optimal linear filter for maximizing the signal to noise ratio (SNR) in the presence of additive stochastic noise. Its high rejection to the noise makes it well suitable to operate with the noisy DIDSON image. Design of multidimensional matched filters was discussed, for example, in [10]. Similarly to the 1-D case, the multidimensional matched filter impulse response is found to be identical with the reflected and delayed multidimensional input signal, and is the classical problem of maximizing the SNR.

The matching filtering is the most time consuming step of the entire process. It is performed using the Intel Math Kernel Library [11], a highly optimized, extensively threaded math routines for scientific, engineering, and financial applications that require maximum performance.

Since the target is a submerged platform in a fixed-Earth position, we assumed to have a-priori knowledge about its orientation with respect to the Earth. With this hypothesis, the filtering operation involves a bi-dimensional convolution:

$$\begin{aligned} C(i,j) &= \sum_{m=0}^{M_a-1}\sum_{n=0}^{N_a-1} A(m,n)\,B(i-m,j-n) \\ 0 &\le i \le M_a+M_b-1 \\ 0 &\le j \le N_a+N_b-1 \end{aligned} \tag{5.4}$$

Table 5.1 Benchmarks of the matching filter

Dimension	Image size	Model size	Time (s)
2	512×96	64×120	0.67
2	512×48	162×110	1.50
3	512×96	64×120	9.00

where A and B are, respectively, the acquired image and the generated image of the target (the model); M and N are the numbers of rows and columns, respectively; and $C(i, j)$ is the resulting convolutions. Both the sonar and the model images used in the convolution are mapped in Cartesian coordinates using Eq. (5.3), since in this representation the model size and shape are less dependent on the distance from the sonar.

For a polar image of 512×48 pixels (DIDSON in low-frequency mode) and a model of 162×110, the computation time was about 1500 m s, on an Intel Pentium-M 1.8 GHz architecture entirely dedicated to execute this algorithm only. This computation time was sufficient to perform an on-line search of the platform during a vehicle sweep. A video of this search experiment is archived on the SAUVIM web site[1] (SD029).

We also tested the case when the orientation of the platform is unknown, by adding another dimension to the convolution. The use of the MKL made it possible to perform this operation in about 9 s. Table 5.1 shows a summary of some of our benchmarks. While the computation times of the bi-dimensional cases (at different DIDSON resolutions) are acceptable to perform online searches (see previously cited video), a matching filter in three dimensions is a bit too heavy for the actual hardware platform. However, it can be considered for off-line algorithms or slow-moving scans.

5.2.4 Localization and Iteration

If a match of the model is found in the current sonar image, the actual location of the target can be computed using the following iterative procedure. Indicating ${}^{0}_{S}\boldsymbol{T}_k$ the transformation matrix of the vehicle frame $\langle S \rangle$ w.r.t. the main frame $\langle 0 \rangle$ and ${}^{S}_{D}\boldsymbol{T}_k$ the transformation matrix of the DIDSON frame $\langle D \rangle$ w.r.t. the vehicle frame, the Cartesian position of the platform at the generic iteration k in the main frame is computed as:

$$\begin{bmatrix} {}^{0}x_{pk} \\ {}^{0}y_{pk} \\ {}^{0}z_{pk} \\ 1 \end{bmatrix} = {}^{0}_{S}\boldsymbol{T}_k \, {}^{S}_{D}\boldsymbol{T}_k \begin{bmatrix} x_{ck} \\ y_{ck} \\ {}^{D}z_{k-1} \\ 1 \end{bmatrix} \tag{5.5}$$

[1] http://gmarani.org/sauvim

where x_{ck} and y_{ck} are the coordinates of the location of the model in the sonar image resulting from the convolution process of Eq. (5.4) and ${}^{D}z_{k-1}$ is computed from:

$$\begin{cases} {}^{D}z_k = 0 & k = 0 \\ \begin{bmatrix} {}^{D}x_{p_k} \\ {}^{D}y_{p_k} \\ {}^{D}z_{p_k} \\ 1 \end{bmatrix} = {}^{D}_{S}\boldsymbol{T}_{k+1}{}^{S}_{0}\boldsymbol{T}_{k+1} \begin{bmatrix} {}^{0}x_{p_k} \\ {}^{0}y_{p_k} \\ {}^{0}z_{p_k} \\ 1 \end{bmatrix} & k > 0 \end{cases} \tag{5.6}$$

Note that, for $k = 0$, the initial location of the target is being approximate as if it lies on the plane identified by the axis x_d and y_d of the DIDSON (Fig. 5.3). Instead, for any subsequent instant in time, the new location of the target is computed by using the projection of the target elevation ${}^{0}z_p$, computed at the previous time step, on the DIDSON frame.

The generic transformation matrix of the target frame $\langle P \rangle$ at the generic instant k can be now built by combining together the known parameters (Euler angles) and the computed coordinates:

$${}^{0}_{P}\boldsymbol{T}_k = \begin{bmatrix} & {}^{0}_{P}\boldsymbol{R}\left(\phi_p, \theta_p, \psi_p\right) & & \begin{matrix} {}^{0}x_{p_k} \\ {}^{0}y_{p_k} \\ {}^{0}z_{p_k} \end{matrix} \\ 0 & 0 & 0 & 1 \end{bmatrix} \tag{5.7}$$

where $\left(\phi_p, \theta_p, \psi_p\right)$ are respectively the roll, pitch and yaw angles of the platform frame $\langle P \rangle$, which are known a-priori. This information can be available in case of search for Earth-fixed and known objects. The transformation matrix ${}^{0}_{P}\boldsymbol{T}_k$ is then used back in the first step (A, Model Building) to compute a new model of the target, and to be used in a new convolution process. By performing, during the iterations, some movements around the target (maintaining it confined within the field of view of the DIDSON), Eq. (5.5) converges toward the absolute Cartesian position of the platform (including the unknown elevation).

The result of the matched filter in presence of the target platform is given in Fig. 5.6. The higher peak corresponds to the location of the platform. This location is given in term of 2D coordinates of the DIDSON image. Considering the size of the DIDSON image 48×512 pixels (in low frequency), the size of the area of the block associated to a single pixel at a distance of 10 m is approximately 0.1×0.02 m. This can be computed with simple geometrical considerations in Fig. 5.3. The actual uncertainty associated with the computed location of the peak is closely related to this limit. Also, results are subject to the quality of the DIDON image. In our attempts, the output of the DIDSON was consistently clear and similar to the one shown in Fig. 5.5, regardless of the condition of the surface of the platform (sand, dirt, etc.).

In order to practically quantify the standard deviation of the position error we repeated several times a pre-defined experiment. For each time, the vehicle was

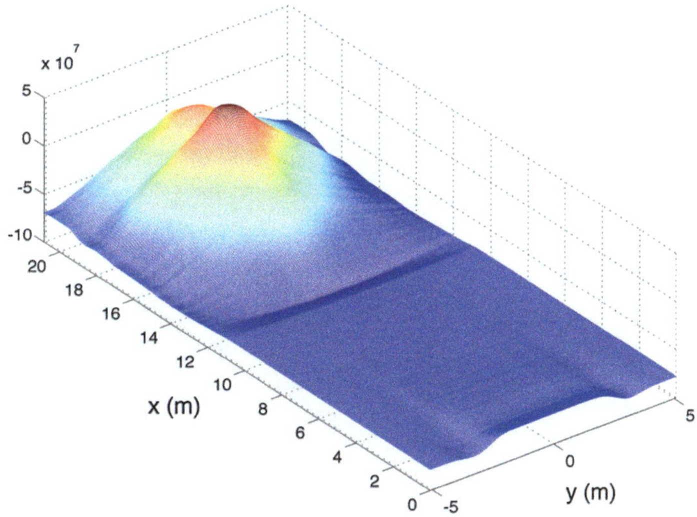

Fig. 5.6 3-dimensional plot of the convolution result

positioned in the same hovering configuration above the platform. In this experiment, the relative Cartesian coordinates of the platform with respect to the vehicle were given by Eq. (5.5), with x_{ck} and y_{ck} computed using DIDSON imagery taken at a distance of 10 m from the target. The bottom facing video camera of SAUVIM was sending back images of a known dimension feature, shown in Fig. 5.7, where the reference red line is 75 mm in length.

Figure 5.7 shows two different final configurations, with only a small difference in the final hovering positions. Considering that the navigation from the area where the DIDSON images were taken to the hovering configuration was done in dead reckoning (thus introducing some extra errors), this experiment confirms an excellent repeatability and, with the successive trials, the global standard deviation of the position error in the order of a few centimeters.

Finally, this information is made available to the whole system including the navigation controller and the main graphical interface (SAUVIM Explorer) located in the ground mission control environment. Figure 5.8 depicts a snapshot of the mission interface during a search test of the submerged platform. The red rectangle is the result, in real time, of the iteration process. For the SAUVIM missions, the platform represents a submerged docking area of the vehicle, and the precise knowledge of its location is fundamental to the autonomous landing procedure.

The SAUVIM Explorer interface also shows the map of the DIDSON image overlapped to the terrain profile. In case of general imagery, since the DIDSON does not disambiguate the elevation of the target, this operation may lead to approximate results (see also the discussion in [12]). In our case, since we acquire elevation maps of the seafloor with other sensors, the sonar image can unambiguously be mapped to the Cartesian space.

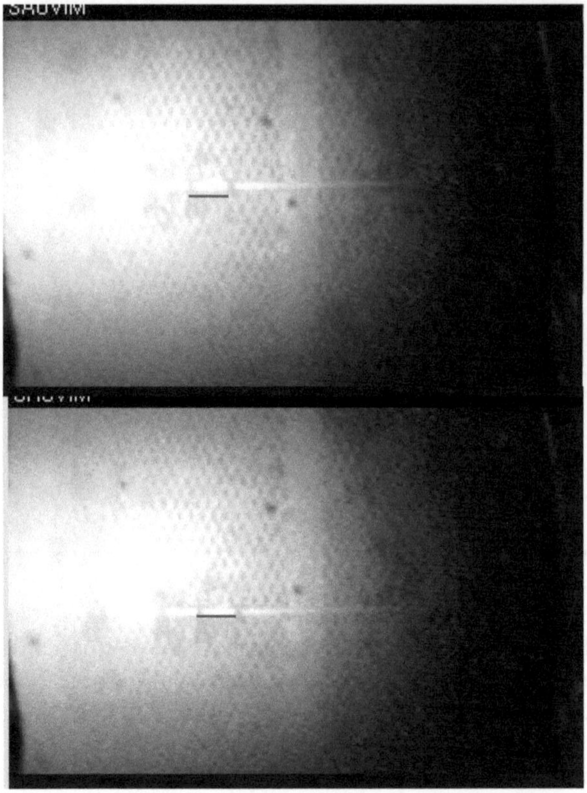

Fig. 5.7 Repeatability of the platform approaching based on the position measurement with the matching filter. The *red mark* is 7.5 cm in length

5.3 Underwater Short-Range Target Localization

At the final stage, when the target is confined within the arm workspace, the localization must be computed with the accuracy necessary to perform the task, for example within a millimeter in case of plugging a connector.

Some sensors, for example video processing and laser or ultrasonic 3D scanners, provide an absolute measurement, even if usually at a low sampling rate and high computational cost. Video processing, however, may present some drawbacks in the ocean depths. The need of a constant light source during the manipulation task may considerably degrade the autonomy of the vehicle. Moreover, the poor visibility in some environments could introduce difficulties in the target detection/recognition process. On the other hand, motion trackers can provide reliable and high sample rate information, but the measurement is relative to the position of the tracking probe. This means that the system must know exactly the relative, generalized position (rotation and orientation) of the probe with respect to the target. However, the sensor probe

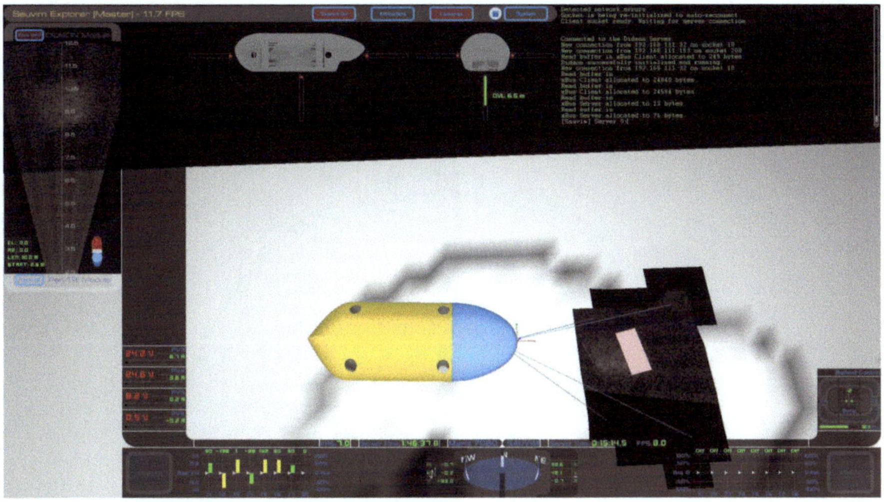

Fig. 5.8 The Sauvim Explorer mission control interface

application/localization has to be done only once and can be achieved substantially in one of the following ways:

- *Operator assisted*. In the case of having a sufficiently reliable link, the application of the sensor probe to the target can be executed by the operator using teleoperation and/or teleprogramming mode [13–15]. This is sometime referred as a semi-autonomous modality of execution of the task.
- *Autonomous mode*. The target localization and sensor probe application are executed in fully autonomous mode with the aid of the above mentioned absolute measurement 3D sensors (camera, scanners, etc.).

After this phase, the manipulation task can be executed using only the information of the motion tracker. The motion tracker-aided manipulation is conceptually similar to the use of a passive arm measurement devices. The main advantage of the motion tracker is the absence of a mechanical link between the target and the AUV, as it uses a simple wire or wireless sensors.

The commercially available motion trackers are mostly developed for virtual reality purposes (for instance in capturing the body movements) or medical use (i.e. for tracking the position of probes).

To validate the feasibility of the ultrasonic tracking during autonomous manipulation, a commercial unit developed for air was modified to work with the robotic manipulator of SAUVIM. This experiment consisted of pouring the content of a test-tube into a container. The system (test-tube seat and container, Fig. 5.9) was prepared with a moving base, whose position and orientation were tracked by the ultrasonic sensor. In Figure 5.9, the little white triangle on the back of the moving base is the ultrasonic tracking sensor. The control software knows the relative generalized

Fig. 5.9 Manipulation with moving target using the ultrasonic tracker

(6 DOF) position of the various objects with respect to the application coordinates, which is sufficient for executing the task.

An underwater version of the 6 DOF tracker was under study, and a preliminary work was presented in [2]. In order to cope with the precision requirements of a generic autonomous underwater manipulation task, it is expected to have the sub-millimeter accuracy, as proven in the preliminary experiments. The underwater tracking technology can also be used in different situations, for example in vehicle docking/undocking procedures [16].

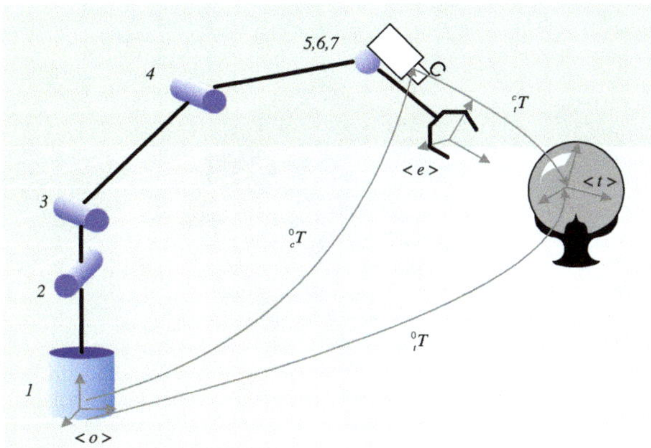

Fig. 5.10 Schematic representation of the arm workcell with camera and target

5.3.1 Localization Using Video Processing

Pose estimation using computer vision is a vast and well established topic. Within SAUVIM, the goal was only to provide the simplest solution for accomplishing the underwater tasks, most importantly enabling workspace optimization introduced in Sect. 4.2.

This goal is achieved using a video camera located on the wrist of the manipulator and a dedicated video processing system [1]. In our task, the target is the dipole as shown in Fig. 4.5. The two spheres composing the dipole are of known diameter and, because its geometrical configuration, it is possible to determine the exact position and orientation with only one camera.

In Fig. 5.10, a sphere is represented schematically with the associated frame $\langle t \rangle$. The transformation matrix ${}^{0}_{t}\boldsymbol{T}$ of the target frame $\langle t \rangle$ with respect to the base frame $\langle 0 \rangle$ is given by:

$$ {}^{0}_{t}\boldsymbol{T} = {}^{0}_{c}\boldsymbol{T}{}^{c}_{t}\boldsymbol{T} \tag{5.8} $$

where ${}^{0}_{c}\boldsymbol{T}$ is the transformation matrix of the camera frame $\langle c \rangle$ with respect to the base frame $\langle 0 \rangle$ and ${}^{c}_{t}\boldsymbol{T}$ is the transformation matrix of the target frame $\langle t \rangle$ with respect to the camera. The placement of the camera ${}^{0}_{c}\boldsymbol{T}$ is easily computed using the joint position information of the arm and the relative position of the camera with respect to the joint on which it is mechanically mounted. This relative position could be precisely computed using a set of predefined movement of the joint hosting the camera along the main axes.

The placement of each sphere location ${}^{c}_{t}\boldsymbol{T}$ with respect to the camera is obtained using video processing of the acquired image. As a matter of fact, the problem may be seen as a 2D localization of a circle within the acquired image. Upon camera calibration, the Cartesian 3D position (x_s, y_s, z_s) of the center of the sphere in the space may be easily computed from the center and the diameter of the 2D circle,

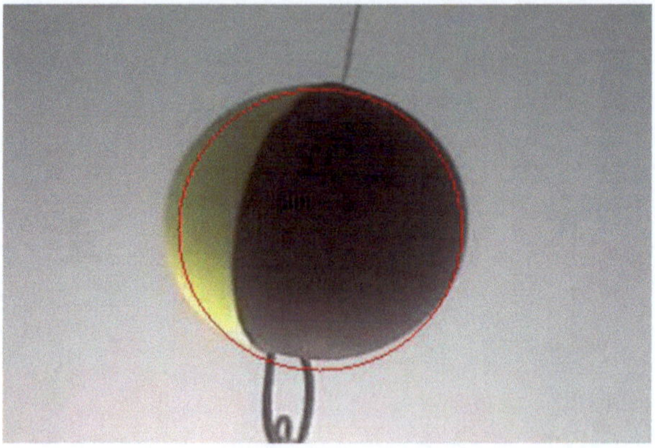

Fig. 5.11 Result of video processing: the circle extraction

using the following relationships:

$$\begin{aligned} z_s &= K_z \frac{r_s}{r_p} \\ x_s &= K_x z_s \left(x_p - x_{p0}\right) \\ y_s &= K_y z_s \left(y_p - y_{p0}\right) \end{aligned} \tag{5.9}$$

where r_s is the actual radius of the sphere, r_p is the measured radius in pixels, x_p and y_p are the position of the center of the circle within the acquired image (in pixels), x_{p0} and y_{p0} are a translation factor depending on the size of the image (in pixels).

The constants K_z, K_x and K_y are the calibration parameters and depend mainly on the focal length of the camera and the physical dimension of the pixels. Their values are computed directly in the water, with the aid of the robotic arm that makes possible the acquisition of a fixed optical pattern from different angles and distances.

The localization of the circle within the image is done using the following sequence of steps:

- Image filtering
- Edge extraction using Canny filter applied to the color image and using the color contrast gradient
- Fast Circle extraction using the line segments found in digital images [17].

This combination of algorithms showed the best performance and robustness with respect to false or missed detections in an underwater environment. Figure 5.11 shows the result of the above sequence of steps applied to a single frame. In our actual implementation, the system is capable of processing about 10 frames per second, which is sufficiently high enough to lock and follow the target when the target moves.

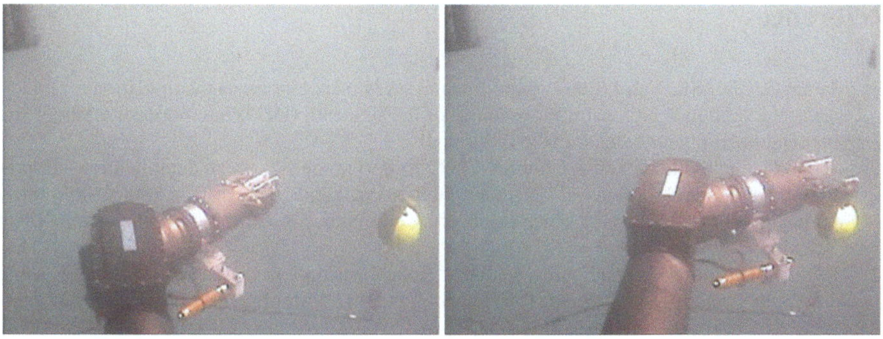

Fig. 5.12 Underwater scenes of the cutting operation

5.3.2 The Cable-Cutting Demo

To examine the effectiveness of the solution presented in Sect. 5.3.1, we considered an underwater mission of locating and cutting a cable by using a special tool integrated with the gripper. The cable passes through a sphere of 10 cm in diameter, which simplifies its localization in the space, and its one end attaches to a floating device. Figure 5.12 shows few snapshots of the operation. The mission was carried out by performing the following tasks in sequence:

- Navigate to the target site.
- Extract the arm and perform a visual scan of the surrounding space, using the attached camera.
- During the scan, try to locate the target (Fig. 5.11).
- Once the target is detected, the arm enters into a tracking state, in order to place the gripper to a constant relative position with respect to the sphere by visual servoing.
- After the tracking system detects no movements of the target for a sufficient amount of time, the arm initiates the cutting task (Fig. 5.12), which is performed in about 2 s with the aid of a special cutting tool integrated into the gripper. During the time frame of the cutting operation, a small movement of the target with respect to the arm may still be corrected using the video feedback.

The only required human intervention is the decision on when to start the visual scan. Once the vehicle reaches the target site, the supervisor confirms that the target is the correct one and then decides to start the autonomous operation. If any error occurs during the above autonomous steps (for example no target was detected during the scan), the arm returns in its initial parking position and the mission is aborted. A mission log allows the operator to verify later the cause of the failure.

References

1. Marani G, Choi SK, Yuh J (2007) Experimental study on autonomous manipulation for underwater intervention vehicles. In: Proceedings of the sixteenth (2007) international offshore and polar engineering conference, Lisbon, Portugal
2. Marani G, Gambella L, Yuh J, Choi SK, (2008) Short range full ocean depth underwater precision 6 dof positon, motion tracker for autonomous manupulation. In: The 18th, (2008) international offshore (ocean) and polar engineering conference & exhibition. Vancouver, British Columbia, Canada
3. Negre A, Pradalier C, Dunbabin M (2008) Robust vision-based underwater homing using self-similar landmarks. J Field Rob 25(6–7):360–377. doi:http://dx.doi.org/10.1002/rob.v25:6/7
4. Briggs A, Scharstein D, Braziunas D, Dima C, Wall P (2000) Mobile robot navigation using self-similar landmarks. In: Proceedings of the IEEE international conference on robotics and automation, 2000, ICRA '00, vol 2, pp 1428–1434. doi:10.1109/ROBOT.2000.844798
5. Belcher E, Hanot W, Burch J (2002) Dual-frequency identification sonar (didson). In: Proceedings of the 2002 international symposium on underwater technology, 2002, pp 187–192. doi:10.1109/UT.2002.1002424
6. Yu SC (2008) Development of real-time acoustic image recognition system using by autonomous marine vehicle. Ocean Eng 35(1):90–105. doi:10.1016/j.oceaneng.2007.07.010, http://www.sciencedirect.com/science/article/B6V4F-4PB0R05-1/2/963242db1af6c4663838ae1ae0184485
7. Yu SC, Kim TW, Marani G, Choi S (2007) Real-time 3d sonar image recognition for underwater vehicles. In: Symposium on underwater technology and workshop on scientific use of submarine cables and related technologies, 2007, pp 142–146. doi:10.1109/UT.2007.370843
8. Walter M, Hover F, Leonard J (2008) Slam for ship hull inspection using exactly sparse extended information filters. In: IEEE international conference on robotics and automation, 2008, ICRA 2008, pp 1463–1470. doi:10.1109/ROBOT.2008.4543408
9. Bradski G, Kaehler A, Pisarevski V (2005) Learning-based computer vision with intel's open source computer vision library. Intel Tech J 9(2):119–130
10. Mastorakis N (1996) Multidimensional matched filters. In: Proceedings of the third IEEE international conference on electronics, circuits, and systems, 1996, ICECS '96, vol 1, pp 467–470. doi:10.1109/ICECS.1996.582907
11. Intel (2013) Intel math kernel library. http://software.intel.com/en-us/node/468334. Accessed 17 Nov 2013
12. Kim K, Intrator N, Neretti N (2004b) Image registration and mosaicing of acoustic camera images. In: Proceedings of the 4th IASTED international conference on visualization, imaging, and image processing, pp 713–718
13. Paul RP, Sayers CP, Stein MR (1993) The theory of teleprogramming. J Rob Soc Jpn 11(6): 14–19
14. Funda J, Lindsay TS, Paul RP (1992) Teleprogramming: toward delay-invariant remote manipulation. Presence: Teleoper Virtual Environ 1(1):29–44. http://dl.acm.org/citation.cfm?id=128947.128950
15. Sayers CP, Paul RP, Catipovic J, Whitcomb L, Yoerger D (1998) Teleprogramming for subsea teleoperation using acoustic communication. IEEE J Oceanic Eng 23(1):60–71
16. Evans J, Redmond P, Plakas C, Hamilton K, Lane D (2003) Autonomous docking for intervention-auvs using sonar and video-based real-time 3d pose estimation. In: Proceedings of OCEANS 2003, vol 4, pp 2201–2210. doi:10.1109/OCEANS.2003.178243
17. Kim E, Haseyama M, Kitajima H (2003) Fast line extraction from digital images using line segments. Syst Comput Jpn 34(10):76–89

Chapter 6
Case Study

Chapters 2 through 5 have introduced all the mathematical details beyond the autonomous manipulation concepts of SAUVIM. This last chapter, as a conclusion of our work, presents the SAUVIM main software framework that allowed us to run all the presented solutions.

Developing the SAUVIM software framework required a considerable effort in integrating together all the components of the system.

In this chapter we also summarize all the remaining technical issues not addressed in the previous chapters, such as motor drivers and path planning.

6.1 The Real-Time Architecture of SAUVIM

The hardware and software architecture of SAUVIM was developed with particular attention to autonomy and global information sharing. It has several similarities to the backseat driver paradigm [1], implemented on a number of platforms (e.g., Bluefin, Hydroid, and Ocean Server). The paradigm refers to a division between "low-level" control and "high-level" control on the vehicle, with most likely the former residing on the vehicle's main computer and the latter on a computer located within a payload section, that can be physically swapped out of the vehicle. The low-level control and the high-level control are also referred to as "vehicle control" and "mission control" respectively. Here, the architecture that coordinates a set of software modules collectively comprising the 'backseat-driver' system running in the payload was implemented by using MOOS (see Newman [2]).

SAUVIM has a precise role separation between high-level (or mission control, in the 'backseat') and low-level (or vehicle control, in the 'front-seat'). This separation was implemented with a dedicated software environment for autonomous systems. The mission control system (backseat) is essentially a software-emulated CPU running a custom programming language, specifically created to simplify high-level operations and algebraic manipulations at the same time. The hardware resides within

G. Marani and J. Yuh, *Introduction to Autonomous Manipulation*,

Springer Tracts in Advanced Robotics 102, DOI: 10.1007/978-3-642-54613-6_6,

an abstraction layer w.r.t. the mission control level. Given a precise and standard specification for the hardware interfaces, this environment can be easily re-adapted to a different hardware layer.

The overall architecture, comprehensive of the above concepts, was structured into a set of layers as schematized in Fig. 6.1. The problem of structuring control functions in a multilayer structure has been studied by others [3, 4]. In the NASREM architecture developed by Albus [3], a theoretical model was proposed consisting of six basic elements: actuators, sensors, sensory processing, world modeling, behavior generation, and value judgment. These elements are integrated into a hierarchical system architecture. Following the same concepts of NASREM, the control system architecture for SAUVIM consists of a set of standard modules and interfaces which facilitates software design, development, validation and test.

The *first* layer, the Very Low Level Controller, is essentially the hardware level.

The *second* one hosts all the software drivers necessary for acquiring the sensor data and sending references to the actuator servo drives.

The *third* layer, the Medium Level Controller, contains the coded algorithms for solving the direct and inverse kinematic problems, as presented in Chap. 3: Resolved Motion Rate Controller, task prioritization, singularities avoidance, collision avoidance and manipulability optimization.

All the input and output of the above algorithms, along with the sensor data from the layer 2, are directly accessible through the FastBus from the *fourth* layer, the High Level Controller. The latter contains our SAUVIM Programming Language (SPL) server, which is in turn connected with the communication layer that allows exchanging its data with the SPL clients (e.g. loading executables). The FastBus, similarly to the global memory in the NASREM model, is responsible for sharing of information between all the layers.

All input and output quantities are accessible from all the modules at all levels and, through the *xBus*. This facilitates the interaction with the user and allows debugging the system.

The following subsections describe in detail the layers of the SAUVIM software and hardware architecture, shown in Figs. 6.1 and 6.2.

6.1.1 Layer 0: The Hardware

The hardware architecture of SAUVIM is a multi-processor system based on a 6U VME bus and the VxWorks real-time operating system. Currently the system consists of:

- Two VME-based computers, hosting the navigation and the manipulator control systems
- Two PC104+ computers dedicated to sensor data acquisition, processing and sharing

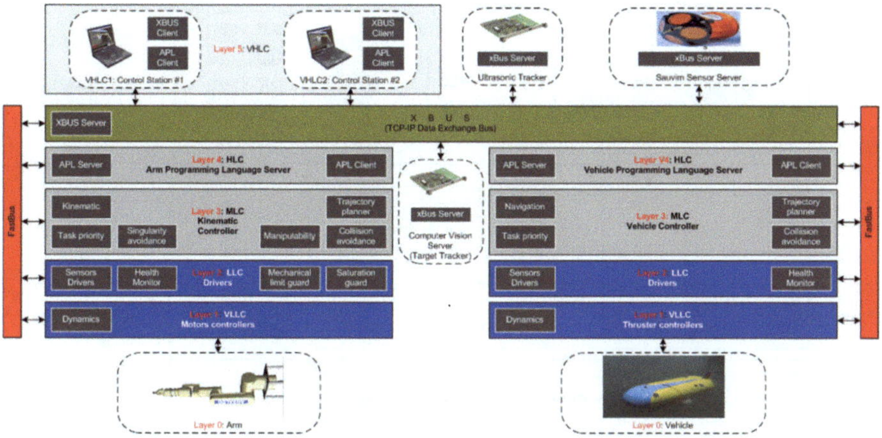

Fig. 6.1 Schematization of the SAUVIM real-time architecture

- One PC104+ hosting the video processing algorithms for the target detection and tracking system
- One PC104+ for the ultrasonic tracker (in development)

The navigation control system includes one main VME CPU board (Motorola), a VME digital/analog I/O board. The two Intel-based PC104+ supplemental boards provide assistance in sensor data processing and collision avoidance computation. All the boards share data through Ethernet-based custom protocols. The navigation control system handles the communication, supervision, planning, low-level control, self-diagnostics, video imaging, etc.

The manipulator control system shown in Fig. 6.3 hosts the second VME-based computer (Motorola) and several hardware-dedicated I/O board. One PC104 board (shown in Fig. 6.2) aids the manipulator control system in the video processing operation necessary to detect and track the target. Data resulting from the video processing subsystem are shared with the whole SAUVIM system (including the SAUVIM Explorer interface), using the same Ethernet protocol (Fig. 6.2).

6.1.2 Layer 1 and 2: Low Level Interface for Robotic Actuators

Many commercial manipulator systems are today equipped with their own complete control system, usually capable of performing tele-operated and programmed trajectories in both the joint and Cartesian spaces. Often, the control system is also accessible via different interfaces (CANopen, EtherCAT, StarFabric, etc.) that allow setting in real time the controller reference input (position and velocity being the most common). In the general control scheme of Fig. 3.2, the above proprietary control system would be enclosed in the *Robot HW Controller* block.

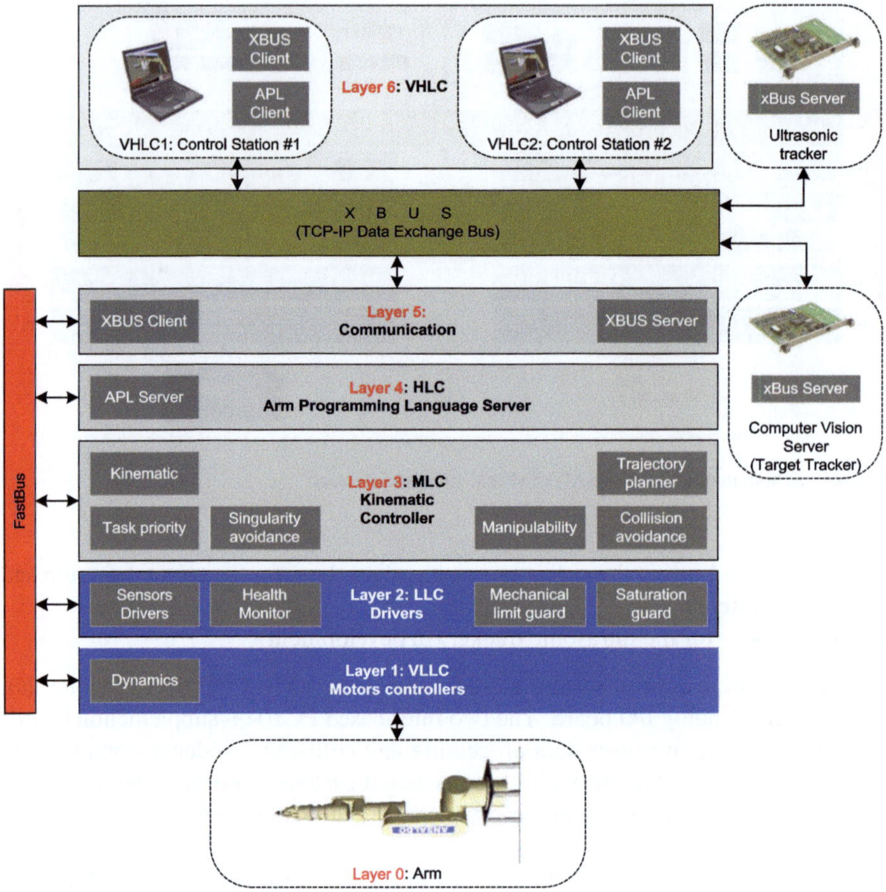

Fig. 6.2 Particular of the manipulator subsystem architecture

As noted in Sect. 3.1, the joint-level control is often realized by dedicated PID control loops, tuned for a wide case of scenarios, and implemented into the *Servo Drive* module. On top of this joint controller, the manufacturer usually implements its own subsystem that may act as a more complex controller (i.e. position, velocity). The latter may not be the best solution when dealing with custom implementations like SAUVIM. Also, the presence of an intermediate control layer introduces inevitably a delay that can be extremely negative in case of compliance (force) control.

When the user is responsible for developing the control strategy, like SAUVIM, it is preferred accessing the actuator in its most natural form. In case of an electrical servo motor driving the robot joint, the natural input is *torque*. This is a direct consequence of the operating principle of an electric motor, where there is a direct relationship between the input current and the generated torque: the latter can hence be generated without a control loop. Controlling velocity or position at the joint

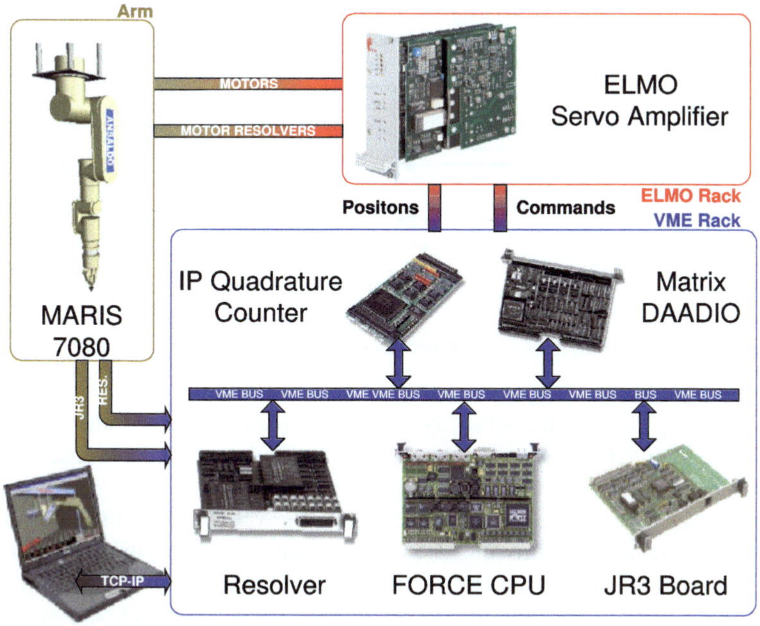

Fig. 6.3 The hardware architecture of the manipulator MARIS 7080

level, instead, is generally achieved by closing the loop with other sensors, and it is realized at the hardware level via PID controller, tuned by the manufacturer for standard conditions or delivered with auto-tuning capabilities.

With the torque servo drive, another important advantage is the considerable reduction of delay. For example, many modern servo drives allow implementing a torque loop sampling rate of several kilohertz. On the contrary, a fully featured manufacturer-supplied control system may introduce delays up to tenth of milliseconds, making the implementation of compliance control with external loops truly problematic.

With a direct access to the servo drive torque input, it is then possible to use Eq. (2.190) to realize a dynamic controller. For example, the control input defined in Eq. (4.33) realizes a well known computed torque controller. The latter sets the dynamics of the position and velocity errors, assuming that the dynamic parameters of the manipulator are known. Within the Lagrange model, it is easy to adapt in real-time to different working conditions, like gravity forces plus buoyancy in water or payloads on the end-effector. Instead, a pre-tuned PID loop in the servo drive may respond differently, depending on the working conditions. Note that many of the above considerations may not be an issue for industrial manipulators, whose working conditions are always same.

The SAUVIM manipulator concept follows the above considerations. The only difference is that, at this stage, we use the internal PID velocity loop of the MARIS

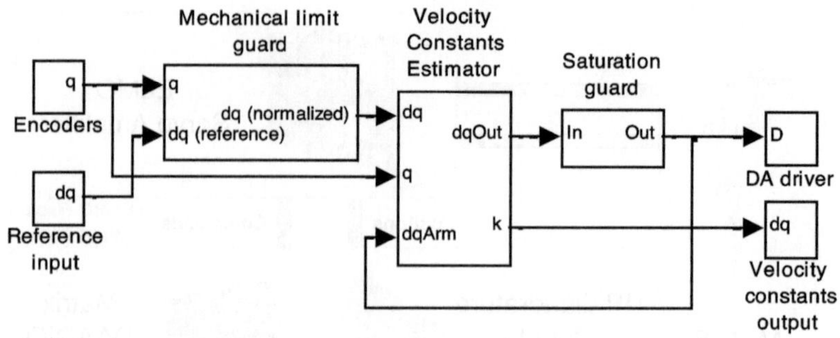

Fig. 6.4 Block diagram of the motors driver module (joint section)

servo drives, by entering the servo drives with a velocity reference (the "Commands" signal line in Fig. 6.3). As noted in Sect. 3.1, this approach provided acceptable performances for testing the task reconstruction algorithm.

The layer 1 in Fig. 6.1 must hence translate the velocity reference into a proper signal for the servo drive module. This was accomplished by pushing the reference through a sequence of signal conditioning blocks in the *Motors driver* module.

The motors driver module, in general, allows driving each actuator according to the relation:

$$\omega_i \text{ (rad/s)} = \overline{k_v}\,\omega_{iref} \tag{6.1}$$

where ω_i is the actual angular velocity of the i-th joint and $\overline{k_v}$ is a constant velocity (typically unitary). Relation (6.1) is valid for each joint except the gripper.

As shown in Fig. 6.4, the Motors Driver module consists of three main blocks:

1. The Mechanical Limit Guard, a software protection for each joint in order to avoid any possible collision against the mechanical stop.
2. The Velocity Constants Estimator, which allows to estimate the velocity constant of each motor controller in order to realize Eq. (6.1).
3. The Saturation Guard, necessary to maintain the required linearity when a motor is approaching its saturation limit.

These blocks are described in the following subsections.

6.1.2.1 Mechanical Limit Guard

This block is responsible for scaling the whole velocity reference vector when a joint approaches its mechanical limit. Internally, the block estimates the time of impact and tries to avoid it. If t_{ti}, the *time-to-impact*, is less than a predefined value $t_{\max}$, the vector of velocities is scaled by a function of the same t_{ti}. Analytically we have:

$$v_{i\,out} = \begin{cases} 0 & q_i \notin \left[q_{i\,min}, q_{i\,max}\right] \\ v_{i\,corr} & q_i \in \left[q_{i\,min}, q_{i\,max}\right] \end{cases} \tag{6.2}$$

where:

$$v_{i\ corr} = \begin{cases} \left(\frac{t_{ti}}{t_{max}}\right)^2 \left(3 - 2\frac{t_{ti}}{t_{max}}\right) v_{i\ in} & t_{ti} < t_{max} \\ v_{i\ in} & t_{ti} \geq t_{max} \end{cases} \tag{6.3}$$

The time-to-impact is estimated as follows:

$$t_{ti} = \min_i \frac{\Delta q_i}{v_{i\ in}} \tag{6.4}$$

with:

$$\Delta q_i = \begin{cases} q_{i\ min} - q_i & v_{i\ in} < 0 \\ q_{i\ max} - q_i & v_{i\ in} \geq 0 \end{cases} \tag{6.5}$$

The vectors:

$$\boldsymbol{q}_{min} = \left[q_{1\ min}, q_{2\ min} \ldots q_{7\ min}\right]^T$$

and

$$\boldsymbol{q}_{max} = \left[q_{1\ \max}, q_{2\ \max} \ldots q_{7\ \max}\right]^T$$

include the values of the minimum and maximum joints mechanical limits.

6.1.2.2 Velocity Constants Estimator

As noted earlier, the servo drives implement a hardware velocity loop PID controller. The angular velocity of each motor is related to the board control input voltage V_{iref}:

$$\omega_i(t)\,(\text{rad/s}) = k_{vi}(t)\, V_{iref}(t) \tag{6.6}$$

where k_{vi} is the board velocity constant. Note that the main difference from Eq. (6.1) is that the velocity constant is different for each board and can be slowly time-variant. Actually it depends mainly on the gear ratio and on the tunable constants of the hardware PID controller.

In order to have a linear behavior as in Eq. (6.1), we first need to estimate all the constants. At the given time instant k, from Eq. (6.6) we have:

$$(k_i)_k = \frac{(\omega_i)_k}{\left(V_{iref}\right)_k} \cong \frac{(q_i)_{k+1} - (q_i)_k}{\Delta t} \frac{1}{\left(V_{iref}\right)_k} \quad [\text{rad}/\left(\text{s} \cdot \text{V}\right)] \tag{6.7}$$

or, similarly:

$$(k_i)_{k-1} = \frac{(\omega_i)_{k-1}}{\left(V_{iref}\right)_{k-1}} \cong \frac{(q_i)_k - (q_i)_{k-1}}{\Delta t} \frac{1}{\left(V_{iref}\right)_{k-1}} \quad [\text{rad}/\left(\text{s} \cdot \text{V}\right)] \tag{6.8}$$

Taking a mean value over a window of n samples, we have:

$$\overline{k_i} = \sum_{j=1}^{n} (k_i)_{k-j} \tag{6.9}$$

Thus, from Eq. (6.1), we have:

$$V_{iref} = \frac{\overline{k_v}}{\overline{k_i}} \omega_{iref} \tag{6.10}$$

In other words, given a reference value ω_{iref}, to have the actual angular velocity of the joint as shown in Eq. (6.1), the voltage reference input to the i-th servo drive must be as shown in Eq. (6.10), and this is the correction performed by the block *Velocity Constants Estimator*.

In the actual implementation, to compute the mean value in Eq. (6.9), only the samples included in a predefined range are considered. Values not satisfying the following condition:

$$\left| k_i - \overline{k_i} \right| \leq \widetilde{k_i} \tag{6.11}$$

are discarded.

$\widetilde{k_i}$ is the (fixed) nominal mean value of the i-th board-joint pair. Note that Eq. (6.11) allows to consider the value 0 as a valid sample, reflecting a failure of the motor. In this case, Eq. (6.10) cannot be used to compute the voltage reference. Instead, if a motor breaks down, we use the a-priori known mean value constant $\widetilde{k_i}$ as the current one:

$$\overline{k_i} = \begin{cases} \sum_{j=1}^{n} (k_i)_{k-j} & \sum_{j=1}^{n} (k_i)_{k-j} > \frac{\widetilde{k_i}}{2} \\ \widetilde{k_i} & \sum_{j=1}^{n} (k_i)_{k-j} \leq \frac{\widetilde{k_i}}{2} \end{cases} \tag{6.12}$$

With this formulation, it is assumed that a motor breaks down when the absolute value of its integrative constant is lower than the threshold $\frac{\widetilde{k_i}}{2}$.

6.1.2.3 DA Driver

This block maps the input integer(s), between 0 and 4095, to the output voltage(s) (between -10 and 9.9951171875 V) of the DAC board:

$$V_{out} = 20\frac{D - 2048}{4096}, \quad 0 \leq D \leq 4095. \tag{6.13}$$

6.1.2.4 Motor Saturation

The motor velocity is limited by the value corresponding to the maximum input voltage allowed by the servo drive board. Because this saturation may affect each joint in a different way, the manipulator may fail to track a reference velocity input. The *SatGuard* blocks ensures a correct tracking by attenuating, proportionally, all the joint velocities when one is approaching its saturation value. Let $\dot{q}_{max}$, the maximum absolute values of the reference components $\dot{q}_i$, be:

$$\dot{q}_{max} = \max_i |q_i| \tag{6.14}$$

The normalized value of each component is given by:

$$\dot{q}_{in} = \begin{cases} \dot{q}_i & \dot{q}_{max} \leq \dot{q}_{lim} \\ \dot{q}_i \frac{\dot{q}_{lim}}{\dot{q}_{max}} & \dot{q}_{max} > \dot{q}_{lim} \end{cases} \tag{6.15}$$

where $\dot{q}_{lim}$ is the maximum value allowed for $\dot{q}$.

6.1.3 Layer 3: Medium Level Controller

The third layer, the Medium Level Controller, is a key component of the system and implements all the mathematical solutions:

- Kinematics (Resolved motion rate controller)
- Task priority (Position prior orientation)
- Singularities avoidance
- Collision avoidance
- Manipulability optimization
- Trajectory planner
- Workspace optimization

As introduced in Sect. 1.1, developing autonomous manipulation increases the level of information exchanged between the system and the human supervisor. In this view, the interface between the task execution layer (layer 4, see Sect. 6.1.4) and the motion controller should be designed to be as simple as possible, for decoupling any low-level controlling issue from the higher level task execution layer.

For example, a simple movement of the end-effector within the arm workspace is executed with the simple grammar:

```
MoveTo(TargetPosition);
```

where `TargetPosition` is a 6-by-1 matrix vector containing the generalized target position (the three Euler angles *roll, pitch, yaw* and the Cartesian positions *x, y, z*.

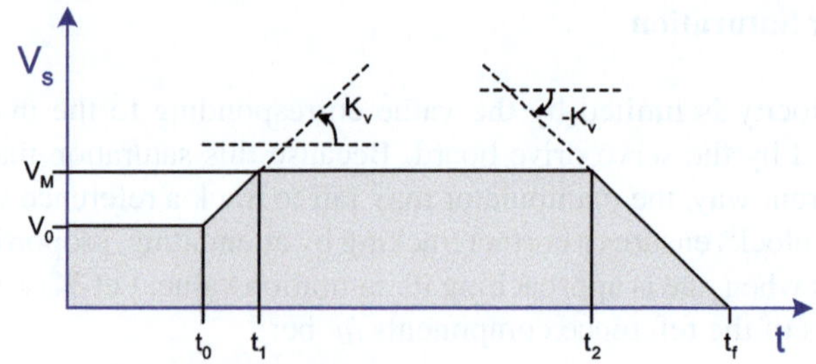

Fig. 6.5 Trajectory generator: trapezoidal velocity profile

The simplicity of the above grammar represents a considerable increase of the level of information exchanged with the robot. The medium level controller is now responsible for ensuring a reliable behavior within the workspace by addressing all the problems mentioned above.

6.1.3.1 Trajectory Planner: A Task-Space Smooth Path Generation

When issuing a simple command like `MoveTo(TargetPosition)`, the medium level controller is also responsible for the computation of the manipulator trajectory.

In general, one of the operation modes of the SAUVIM control software is *Tracking*, which allows driving the end-effector through a predefined path. The path is defined by a succession of points X_i, each one representing a generalized position in the task space:

$$\boldsymbol{X}_{ei} = \begin{bmatrix} roll_i \; pitch_i \; yaw_i \; x_i \; y_i \; z_i \end{bmatrix}^T \tag{6.16}$$

The rotation parameters indicate the orientation of the end-effector with respect the robot main frame. The trajectory between $\boldsymbol{X}_i$ and $\boldsymbol{X}_{i+1}$ is planned for preserving the continuity of the task velocity and avoiding excessive stress to the arm structure.

Figure 6.5 illustrates a one-dimensional velocity profile that accomplishes the above requirements.

Given the input data for the profile as s_0 (the initial position), s_f (final position), V_{s0} (initial velocity), V_m (maximum velocity), K_v (acceleration) and t_0, the output trajectory is:

$$s(t) = \begin{cases} s_0 & t = t_0 \\ s_0 + \int_{t_0}^{t} V_0 + K_v \cdot (t - t_0) dt & t_0 \le t < t_1 \\ s_1 + \int_{t_1}^{t} V_m dt & t_1 \le t < t_2 \\ s_2 + \int_{t_2}^{t} V_M - K_v (t - t_2) dt & t_2 \le t < t_f \\ s_f & t \ge t_f \end{cases} \tag{6.17}$$

where:

$$s_1 = s_0 + \int_{t_0}^{t_1} V_0 + K_v \cdot (t - t_0)dt \tag{6.18}$$

$$s_2 = s_1 + \int_{t_1}^{t_2} V_m dt \tag{6.19}$$

$$s_f = s_2 + \int_{t_2}^{t_f} V_M - K_v(t - t_2)dt \tag{6.20}$$

Expanding above integrals we have:

$$s(t) = \begin{cases} s_0 & t = t_0 \\ s_0 + \frac{1}{2}K_v \cdot \left(t^2 - t_0^2\right) + V_0 \cdot (t - t_0) - K_v t_0 \cdot (t - t_0) & t_0 \le t < t_1 \\ s_0 + \frac{1}{2}K_v \cdot \left(t^2 - t_0^2\right) + V_0 \cdot (t_1 - t_0) - K_v t_0 \cdot (t_1 - t_0) \\ \quad + V_m \cdot (t - t_1) & t_1 \le t < t_2 \\ s_0 + \frac{1}{2}K_v \cdot \left(t_1^2 - t_0^2\right) V_0 \cdot (t_1 - t_0) - K_v t_0 \cdot (t_1 - t_0) \\ \quad + V_m \cdot (t_2 - t_1) - \frac{1}{2}K_v \cdot \left(t^2 - t_2^2\right) + V_m \cdot (t - t_2) & t_2 \le t < t_f \\ \quad + K_v t_2 \cdot (t - t_2) \\ s_f & t \ge t_f \end{cases} \tag{6.21}$$

where:

$$t_1 = t_0 + \frac{V_m - V_o}{K_v} \tag{6.22}$$

$$t_2 = t_f - \frac{V_m}{K_v} \tag{6.23}$$

$$t_f = \frac{2s_f K_v - 2s_0 K_v + 2t_0 K_v V_m + 2V_m^2 - 2V_m V_0 + V_0^2}{2K_v V_m} \tag{6.24}$$

With a particular set of input parameters, it is possible that $t_2 \le t_1$. In this case, the trapezoidal profile of Fig. 6.5 degenerates into the triangular profile of Fig. 6.6, because the above condition ensures that maximum velocity will be smaller than V_m.

In this case we have:

$$V_{se} = \frac{\left(t_f - t_0\right) K_v + V_{s0}}{2} \tag{6.25}$$

$$t_e = \frac{t_f + t_0}{2} - \frac{V_{s0}}{2K_v} \tag{6.26}$$

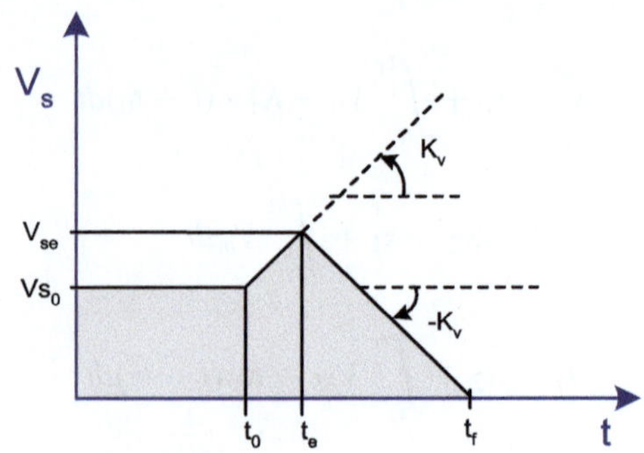

Fig. 6.6 Trajectory generator: triangular velocity profile

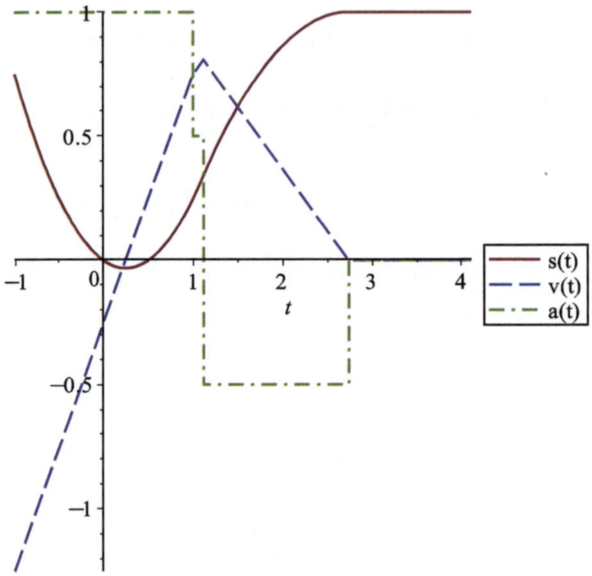

Fig. 6.7 Trajectory generator example: position $s(t)$, velocity $v(t)$ and acceleration $a(t)$ in the triangular case, with $t_0 = 1$, $s_0 = 0.25$, $s_f = 1$, $V_{s0} = 0.75$ and $K_v = 0.5$

$$t_f = \frac{-V_{s0} + t_0 K_v + \sqrt{2V_{s0} + 4s_f K_v - 4s_0 K_v}}{K_v} \tag{6.27}$$

Figures 6.7 and 6.8 show two numerical examples, for one-dimensional case.

The extension to the multidimensional vector (6.16) is done with the following considerations.

Let $\boldsymbol{R}_0$ be the initial rotation matrix of the end-effector with respect the main frame. Likewise, let $\boldsymbol{R}_f$ be the final posture. It is possible to find a direction $\boldsymbol{r}_i$ around which $\boldsymbol{R}_0$ must be rotated of θ radians to obtain $\boldsymbol{R}_f$. Thus we can define the four-element vector as:

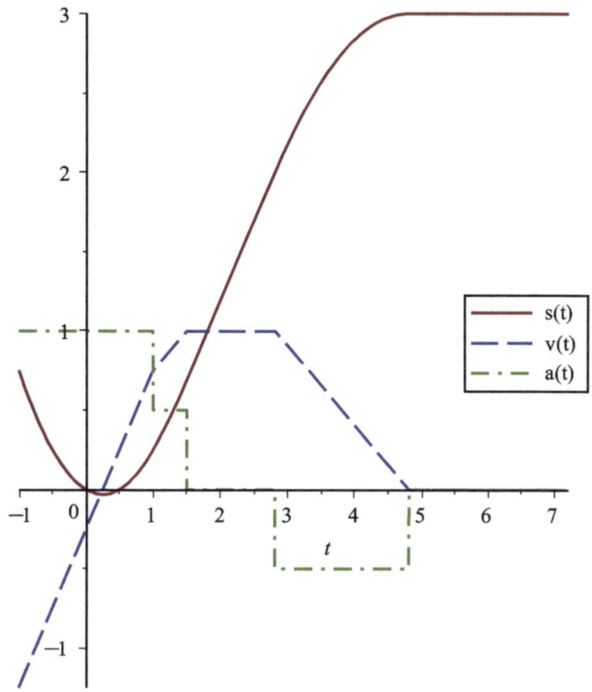

Fig. 6.8 Trajectory generator example: position $s(t)$, velocity $v(t)$ and acceleration $a(t)$ in the trapezoidal case, with $t_0 = 1$, $s_0 = 0.25$, $s_f = 3$, $V_{s0} = 0.75$ and $K_v = 0.5$

$$AB_0 = \begin{bmatrix} x_0 \\ y_0 \\ z_0 \\ 0 \end{bmatrix}, \quad AB_f = \begin{bmatrix} x_f \\ y_f \\ z_f \\ \theta \end{bmatrix}, \tag{6.28}$$

and the parameter:

$$s_f = \left|AB_f - AB_0\right| \tag{6.29}$$

obtaining:

$$AB(t) = AB_0 + \left(AB_f - AB_0\right)\frac{s(t)}{s_f} \tag{6.30}$$

where $s(t)$ is obtained by integrating the trapezoidal or triangular profile as above. The rotation matrix $\boldsymbol{R}(t)$ can be computed from the fourth component of $AB(t)$, say $\theta(t)$, by rotating $\boldsymbol{R}_0$ around $\boldsymbol{r}_i$ of $\theta(t)$ radians.

Finally, for generating a path different from a straight line (i.e. circles, etc.), the solution is to represent the multi-dimensional curve with parametric equations, and use the trapezoidal profile for generating the evolution of the mono-dimensional curve parameter.

Please note that the relationship between the task acceleration and joint acceleration has not been derived. As a precaution, in order to physically limit the acceleration on each joint, especially during the turn-on and turn-off stages, the velocity control signal is filtered by a rate limiter. The upper limit of the joint velocity rate is high enough not to interfere with the task linearity, yet avoiding dangerous joint accelerations.

The same path generation algorithm is applied to the navigation controller. In this case, the maximum accelerations are chosen within the limits imposed by the vehicle thrusters.

6.1.4 Layer 4: High-Level Robot Programming Language

The choice of an appropriate programming language must address a wide range of issues. Many autonomous systems operate in an unstructured and dynamic environment. Here, the language must have the necessary flexibility to react to that world using sensor information and the available actuators. The unpredictability of events requires that a generic control algorithm may be interrupted at any time, in order to face inconsistent or incomprehensible inputs and preserve the safety of the system.

A generic language must have the capability of interacting with the hardware via an abstraction layer. For example, a robot drawing on a white board should accept inputs like pick-the-pen, find-the-board, draw-letter, erase-the-board and so on. Because each of the above operations requires complex low-level behaviors, a good robot programming language should span different layers. The interaction between the hardware and the programming language is one of the major problems in attempting to universally unify a generic robot programming system. Many companies developed their own language suitable for a particular system, hence the fact that no uniform consensus has been given to a particular language, is not surprising.

Another very important issue that a robot programming language must address is the time interaction. A generic control system is usually hosted by a real-time operating system, spawning several periodic tasks running at a fixed sampling time. A precise real-time interaction with the low level control system enables the high level programming language to correctly quantify discrete-time algorithms (e.g. integrators, derivators, etc.) or to perform additional high-level operation like real-time tracking of time-dependent trajectories. The layer of the language where part of the control algorithm resides must have the capability of synchronizing with the above sampling time, monitoring the execution length to avoid exceeding the time-limit.

Finally, for a large class of autonomous systems like underwater robots, it is necessary to organize the language subsystems in a client-server architecture, in order to separate the human interface from the execution layer. In fact, they may have to reside in separate environments (e.g. the vehicle and the ground station).

Within the SAUVIM project, we introduced a new robot programming language, developed with a careful consideration of all the above issues. The SAUVIM Programming Language (SPL) is completely math-oriented and includes a flexible hardware abstraction layer, capable of interfacing with a variety of sensors and actuators. SPL uses a client/server model, and can be manipulated on any computer with network connectivity to the robot. In addition, SPL supports multiple concurrent client connections to devices, creating new possibilities for distributed and collaborative sensing and control.

6.1.4.1 Summary of Robot Developing Environments

Text-based control-specific languages are the most common method of controlling industrial robots. An extensive review and classification has been done as shown in [5–7]. Most of these programming languages are very simple with a BASIC-like syntax and simple commands for controlling robot behaviors. Their biggest problem is the lack of a universal standard from robot manufacturer. Often robot manufacturers also provide a simulation environment.

The recent works in text-based systems have diverged from these robot-specific languages to develop more general purpose high-level programming languages suitable for any robot. Typically, this involves extending existing languages such as C++ [8], Java [9, 10], and Haskell [11].

Among the newest robot software frameworks, two of considerable attention are OROCOS [12–15] and ROS [16].

Open Robot Control Software (OROCOS)

The Open Robot Control Software (OROCOS) project uses both the component-based and object-oriented reusability strategies. The project has yielded four C++ libraries, two geared toward each strategy. The Real-Time Toolkit (RTT) and OROCOS Component Library (OCL) establish a component-based infrastructure and a library of ready-to-use components, providing the high level management of interactions within an application. Components exchange information primarily through abstracted Data-Flow Ports, an anonymous publish-subscribe system in which a component does not know where its inputs are originating or where its outputs are being utilized. These Data-Flow Ports may be either buffered, in which case messages are stored in a single queue per link, or unbuffered, in which case only the most recent message is available to be read by any subscribers. This decomposition and isolation of the components enables easier specification of the interactions taking place within the software and easier validation of individual components, because this organization is conducive to generating test inputs and examining the resulting outputs. OROCOS includes a scripting language that allows users to write programs and state machines controlling the system in a user-friendly realtime script language. The advantage of scripting is that it is easily extendible and does not need recompilation of the main program. OROCOS is probably one of the most complete robot software frameworks, implementing the majority of the issues discussed above. The only drawback is the difficulty to port the framework to embedded system like the hardware of SAUVIM. For this reason, and for the fact that SAUVIM began before OROCOS was available, our choice was toward supporting the development of a custom framework.

Robot Operating System (ROS)

ROS is an open source robot software platform designed with the primary goal of enabling software reuse. It is intended to be a thin architecture, providing sufficiently a few constraints as to be integrable with software written for other platforms, such as the aforementioned OROCOS and Player [17]. Like OROCOS, it employs both the component-based and object-oriented reusability strategies. ROS provides the infrastructure for writing software components, called nodes, and for exchanging information between these nodes via an anonymous publish-subscribe model.

6.1.4.2 SAUVIM Programming Language

The SAUVIM distributed programming environment for autonomous systems is written in ANSI-compliant C and C++, and can be cross-complied for different platforms (VxWorks, Linux, Unix, Mac OS-X, Windows). This capability enables breaking the environment into separate parts, the software-emulated CPU and the code generator ("complier"): the execution CPU can run inside the real-time controller (for instance running a VxWorks operating system) while the compiler may reside on a remote platform such as Windows or Unix, linked via the communication system. Figure 6.1 shows the case where the remote client is a personal computer residing externally (at least when the communication link with the vehicle is available).

This configuration is duplicated for the manipulator and linked together with the SAUVIM navigation system through the main communication layer xBus (Fig. 6.1).

The reliability of the above modularization was confirmed by a successful remote operation of SAUVIM, carried out in year 2010, when the manipulator was instructed to perform a simple task from West Virginia, with the vehicle standing in the water of the Hawaiian ocean.

Overview of SPL

The SAUVIM Programming Language subsystem was developed using the well-known tools Lex [18] and Yacc [19], a lexical analyzer and parser generator. Like C, Fortran, Pascal, Basic, and so on, SPL is a procedural language, in the sense that it consists of a sequence of commands, which are executed strictly one after another. Similarly to other procedural languages, the code may be organized into procedures and libraries, which simplifies the separation of the high level (task oriented) layer from the mid-level layer. As a matter of fact, the latter consists of a set of procedures for attaining particular behaviors.

SPL is not strongly typed like C and Pascal and no declarations are required. It is more like Basic and Lisp in this respect. However types exist: type checking is done at run time and must be programmed explicitly.

The programming language, although pseudo-compiled, is in turn interpreted by a software-emulated CPU. For this reason it is not suitable for running numerically intensive programs because of the virtual CPU overhead. Therefore, the most computationally expensive algorithms have been implemented in a different layer (the

MLC in Fig. 6.1) and are accessible via the SPL abstraction layer, as better explained in Sect. 6.1.4.2.

Another important feature of the SAUVIM Programming Language is the parallelization of servers. This option allows spawning different "processes" running simultaneously with the possibility of mutual interruption in case of particular events. The parallelization is very important when running, for example, health-monitoring procedures or for handling any kind of exceptions.

Finally the language is math-oriented and offers the possibility of symbolic manipulation of expressions, arrays and/or matrices.

Programming in SPL

The SPL syntax for the assignment is taken from Algol 60. The assignment statement looks like:

```
name := expr;
```

where `expr` is any expression and `name` is a variable name. The main difference between SPL and traditional programming languages is that the generic identifier is generally an entity that, if not assigned, stands for itself. In other words it is a symbol. Symbols are used to represent unknowns in equations, variables, indices, etc. Consider the assignment statement:

```
P := x^2 + 4*x + 4;
```

Here the identifier `P` has been assigned the formula `x^2 + 4x + 4`. The identifier `x` has not been assigned a value: it is just a symbol, an unknown. This assignment automatically sets the type of `P` to expression. The identifier `P` is now like a programming variable, and its value can be used in subsequent calculations just like a normal programming variable.

The conditional statement has the following syntax:

```
if expr then statseq
[ else statseq ]
end if
```

where `statseq` is a sequence of statements separated by semi-colons and `[...]` denotes an optional part. A typical `if` statement would be:

```
if x < 0 then -1; else 1; end if
```

The `for` statement has the following syntax:

```
for name from expr to expr do
    statseq
end do
```

Finally a conditional loop can be constructed according to the following syntax:

```
while expr do
    statseq
end do
```

Table 6.1 SPL data types

null	numeric	name	matrix
intvector	+	*	^
builtinfcn	function	exprseq	stmtseq
executable	coeffterm	unknown	repeat
integer	list	indexed	if
ifelse	bool	asm	lrstack1
while	for	procedure	nameseq
string	uneval	eval	range
<	and	=	not
or			

The loop is executed while the condition `expr` evaluates to true.

Data Types

Although it is not necessary to declare the type, a generic variable belongs to the class type corresponding to its content. Many pre-defined types exist in SPL: a type can be any one listed in Table 6.1. Type-checking can be realized at run-time. For instance, the expressions:

```
type(x+y,"+");
type(1.2345,"string");
```

return respectively `true` and `false`. Types can be structured together in lists, similarly to the struct of C language. The resulting list object is a single entity composed of sub-objects of different types. This is necessary when working, for example, with map, search and map theory.

The most relevant is probably the matrix type. A matrix is a generic 2-D array, which in SPL can be constructed with the following syntax:

```
M := matrix(nRows, nCols,[[R1],[R2],...]);
```

where `nRows` and `nCols` are the dimensions of the array and `Ri`, the i-th row, is a sequence of entries separated by commas. For example, a generic column vector can be defined as:

```
q := matrix(3,1,[[q1],[q2],[q3]]);
```

SPL Procedures

The capabilities of SPL can be customized and extended by using procedures. In general, a SPL procedure has the following syntax:

```
proc (NameSequence)
    [local nameseq;]
    [global nameseq;]
    statseq
end
```

where `nameseq` is a sequence of symbols separated by commas, and `statseq` is a sequence of statements separated by colons or semicolons. The scope of visibility of local variables is within the procedure, while global variable are visible within the main workspace and within all the procedures in all the instantiated parallel servers. This makes possible exchanging data across all the spawned processes.

A procedure definition is a valid expression that can be assigned to a name. As an example, the command below assigns to the symbol `f` an user-defined procedure that adds two arguments `x` and `y`:

```
f := proc(x, y)
    x + y;
end proc;
```

This procedure has two parameters `x` and `y`, no local variables and only one statement. The value returned by the procedure is `x+y`. In general, the value returned by a procedure is the last value computed. The following procedure call evaluates `f` with the arguments `3` and `5`:

```
f(3, 5);
```

SPL allows symbolic manipulation of variable. For example, `f` can be invoked with symbolic input:

```
f(Joint1, Joint2);
```

The Hardware Abstraction Layer

The interaction between the hardware and the programming language is performed by a subset of built-in SPL procedures, fully user-defined in order to meet the requirements of the hardware device drivers.

A built-in procedure for accessing the hardware level is essentially an interface between SPL and any custom C/C++ code. It must be defined within the SPL source code before compiling the programming language system. Currently there are more than 100 built-in procedures and most of them perform some specific action like enabling the motors, getting the sensor information data, sending the motor velocity reference, etc. For example, the following syntax can be used in order to assign the variable `q` with the values of the joint angles:

```
q := GetJointAngles();
```

which returns a 7-by-1 matrix containing the seven joint angles of the manipulator (in radians). Internally, the above procedure executes the following steps:

1. Creates an empty 7-by-1 matrix.
2. Reads the IP-quadrature board via the device driver.
3. Fills up the 7 elements of the matrix with the above values.
4. Returns the matrix.

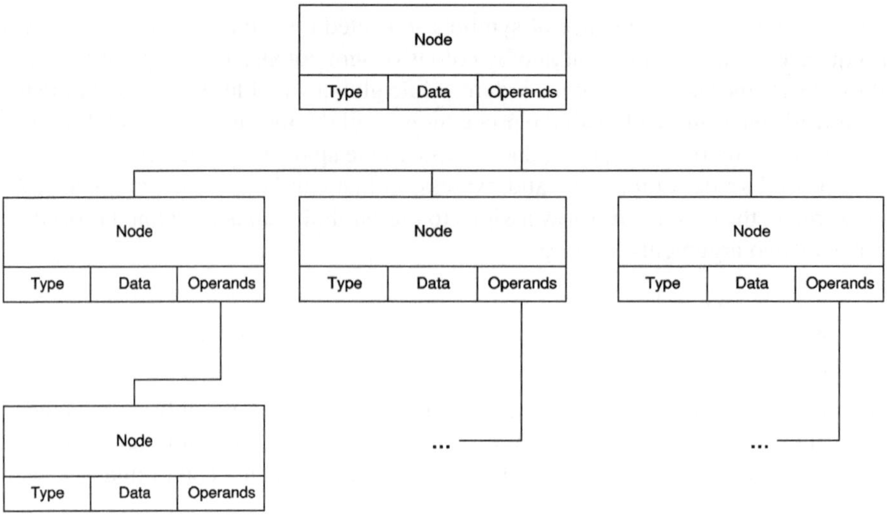

Fig. 6.9 Internal representation of objects: the syntax tree

The internal C++ interface of SPL simplifies the writing process of built-in functions with a large amount of classes, member functions and operators.

Memory Management

Internally, a generic object is represented by a syntax tree of nodes (see Fig. 6.9). Each node has an assigned type, a data field and a list of operands. Data are interpreted according to the type of the node, and internally are stored in a special array with dynamic length and type. For example, the data field of a node of type numeric contains the numeric value of the node (in double-precision), while the operands field is empty. Conversely, a node of type function contains no data and a variable number of operands (the function arguments). A function can be for example the addition operator: in this case the operands are the addends. Each addend can be, in turn, any generic node. This allows the easy handling of expressions in a symbolic form. The C++ interface provides all the necessary support for creating the generic node and manipulating its type, data and operand fields. The creation/destruction process is assisted by a management algorithm which allows optimizing the allocated memory and avoiding memory leaks. Inside SPL, each node is stored in a dynamic array called memory in Fig. 6.10. Each node is referenced by a special class that keeps track of the number of references to the node itself. When the node has no more references, it is marked as free and can be used for new contests. This allows the efficient reuse of the already allocated nodes, keeping the size of the dynamic memory to the indispensable minimum. This is very important when the server is running in embedded systems with limited amount of resources. Figure 6.10 also shows the basic concept that SPL uses for the execution phase. A software-emulated virtual CPU is responsible to fetch and execute the SPL statements according to the following steps:

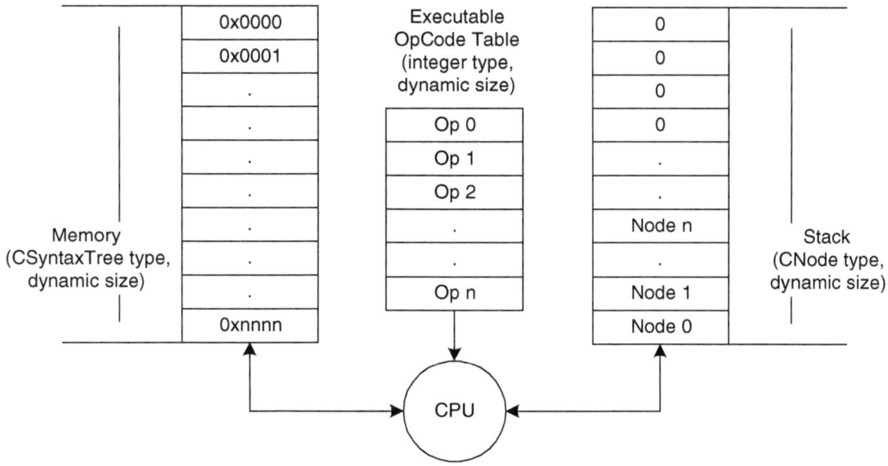

Fig. 6.10 Internal architecture: execution of statements via the software-emulated CPU

1. The executable, consisting of an op-code table and a tree of data nodes, is loaded into the state machine system of Fig. 6.10. More precisely, the op-code table is stored in the *executable table* while a set of data nodes is loaded into the *memory*.
2. The virtual CPU fetches the next available op-code from the current op-code table.
3. The required arguments involved in the execution of the current statement are pushed from the memory to the stack.
4. The CPU executes the current op-code, popping the arguments and pushing the result on the stack.

The above concept is replicated in parallel for each instantiated server. At the end of execution, after the last op-code, the CPU enters an idle status.

Client-Server Architecture

As earlier noted, the architecture of SPL is client-server oriented: the human interface (workspace input) and the execution layer reside in separate environments, communicating via a special protocol over TCP-IP (Fig. 6.11). This allows the accomplishment of one of the main requirements of the NASREM model: with a workspace input, the human operator can take over any layer at any time.

The client is generally a command-line input interface enabling loading libraries, executing statements and/or controlling the execution of the server. Figure 6.12 shows a snapshot of our console implementation.

The operations involved on the client-side are the followings:

1. Preprocess the source code
2. Compile and create the executable
3. Encode the executable
4. Send the encoded executable via TCP-IP to the SPL server

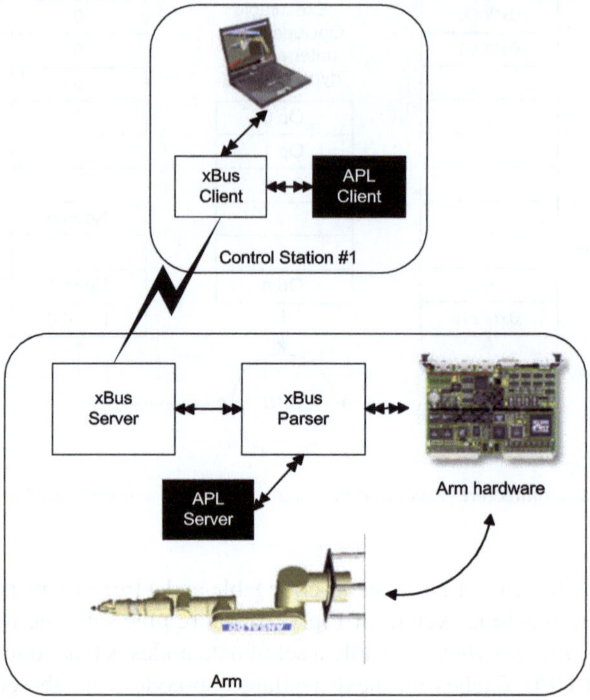

Fig. 6.11 The client-server architecture

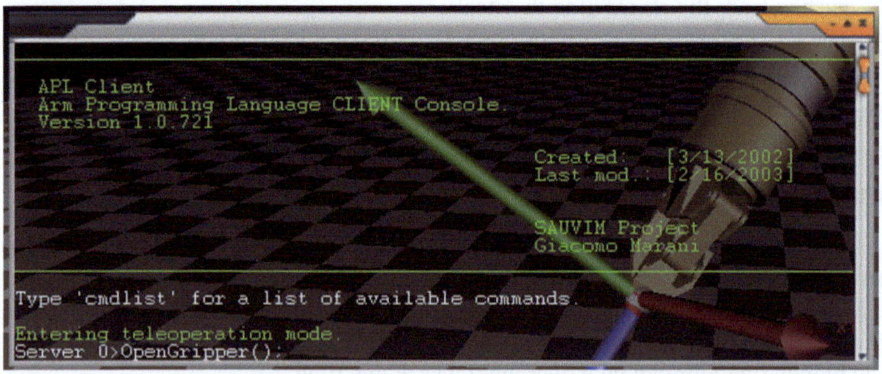

Fig. 6.12 The SPL client console

On the server-side, once the encoded executable has been received, the following operations take place:

1. Decode the executable
2. Load the executable in the virtual CPU
3. Execute

One important feature of the overall programming language system is the possibility of using more than a single client. This is important for monitoring from a secondary client the overall behavior of the primary client, which may reside inside a separate autonomous system. For example, in our autonomous underwater vehicle, the primary client is inside the vehicle CPU, delivering the operation commands to the arm. On the ground side, the user can monitor, with a secondary client, the behavior of the arm and eventually take control over the main CPU.

6.1.5 Layer 5: The Communication Layer

The fifth layer implements a TCP-IP based client-server communication system. The server can handle multiple clients using a robust communication protocol capable of auto-reconnection in case of a temporary network failure. This is important in a hostile environment, where the communication media does not allow safe and durable connections (like acoustic modems).

Internally, the server is a meta-state machine, or a set of finite state machines. These machines, one for each client connection, are sequentially called so that each one can request a different command execution to the parser. It is a matter of the parser allowing or not the execution of each command, according to its priority with respect to the already running ones.

An internal timer generates an error if any reading or writing operation cannot be executed within a certain amount of time. The error results in a forced disconnection followed by a reconnection attempt by the client. Upon any disconnection (graceful or forced), the correspondent server state machine is destroyed and removed from the machine list. Forced disconnections may happen even for other kind of socket errors (for example when losing the carrier of the acoustic modem during a mission), and are always followed by a reconnection attempt. For example, during the execution of any task, it is possible to unplug and successively re-plug the network cable: the only consequence is the loss of control by the client during the time that the cable is disconnected.

The kernels of both server and client are written in Ansi-C language, including its vectorial state machine. This facilitates building the system directly from the source code on a different platform, such as Windows (for the client or a generic simulation server) or VxWorks (the actual arm controller).

6.2 Application Example

In conclusion of our work, we would like to present the SAUVIM final demonstration conducted on January 20, 2010, at Snug Harbor in Honolulu, Hawaii, at the end of the project. The goal of the SAUVIM final demonstration was to demonstrate the capabilities of the overall system, with particular attention to the autonomy

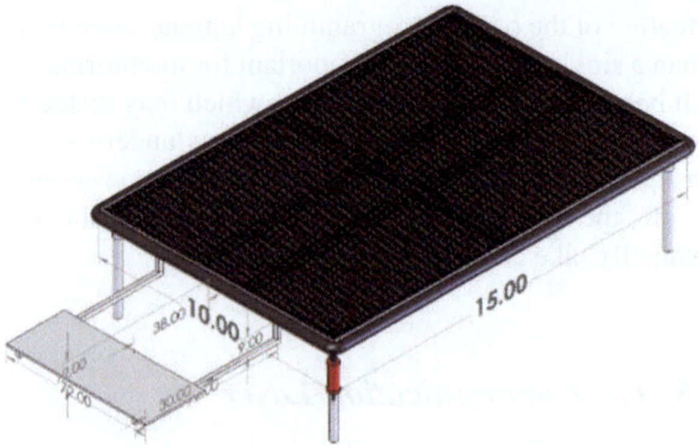

Fig. 6.13 The underwater docking platform

aspect, and including repeatability. This live demonstration confirmed a technological breakthrough in the field, as the autonomous manipulation had been a bottleneck for underwater intervention missions.

A video of the following demonstration is available on the main page of the SAUVIM website.[1]

The demonstration was about an underwater object recovery mission, organized into of the following 6 phases:

- Phase 1: Undock from the pier and navigate to a search area
- Phase 2: Search for the submerged platform
- Phase 3: Navigate and dive toward the platform
- Phase 4: Hover (station keeping)
- Phase 5: Hook a recovery tool to the target object (autonomous manipulation)
- Phase 6: Return to the pier

Each phase was executed with no human intervention. Only the following information was known in advance and given to the SAUVIM computers:

1. Shape of the platform (Fig. 6.13)
2. Shape of the target for recovery (Fig. 6.14)
3. Search area for the platform (Fig. 6.15)
4. Directional information about the target location w.r.t. the platform (frontal area of the platform)

The beginning of the mission consists in launching, from the remote SPL client, the "start-mission" command. This is so far the only human intervention necessary.

Of course, according to the considerations in Sect. 6.1.4.2, it was possible to interrupt the system and gain manual control of the mission at any time, with the ground mission control client.

[1] http://gmarani.org/sauvim

Fig. 6.14 The target to recover

Fig. 6.15 Satellite image of the demonstration area (Snug Harbor, Honolulu, Hawaii). Courtesy: Google

The high-level description of all the subsequent phases was completely coded in SPL, and the excellent flexibility of the programming language in describing an autonomous mission was demonstrated.

6.2.1 Phase 1: Undock from the Pier and Navigate to a Search Area

Figure 6.15 shows a satellite view of the demonstration area. The vehicle is initially docked at the upper-left corner of the area, at top side of the red path.

Fig. 6.16 The beginning of undocking phase

After issuing the "start-mission" command, SAUVIM is programmed to perform surface navigation, aided by the DGPS, and consisting in the following two simple phases:

- Maneuver to undock from the pier
- Reach the center of the harbor

Figure 6.16 shows the vehicle at the beginning of the undocking phase, trying to make its way toward the center of the harbor.

During this phase the arm is inactive, while the vehicle dynamic controller (Sect. 4.3.1.1) is acting in 3 DOF only (surface navigation). In this phase, the PHINS uses the DGPS for its positional measurement input, resulting in an accuracy of about 2-3 meters. This accuracy is sufficient for moving the vehicle to a search area (Fig. 6.15).

6.2.2 Phase 2: Search for the Submerged Platform

After arriving at the search area, SAUVIM begins the maneuvers for locating the platform. The search for the platform is executed with the aid of the DIDSON sonar, following the procedure described in Sect. 5.2.

The vehicle performs a circular scan of the ocean bottom at the maximum SONAR range (40 m), keeping the platform pose identification algorithm active, described in the above cited section. The depth of the ocean in this area is about 10 m, making it possible to collect a detailed scan of the sea bottom. Figure 6.17 shows the output of SAUVIM Eplorer, which reports the real time mapping of the sea floor.

The scan terminates once the platform has been detected by the pose estimation algorithm. From then, the vehicle enters a tracking mode by performing position adjustments until it reaches a predefined relative position with respect to the platform.

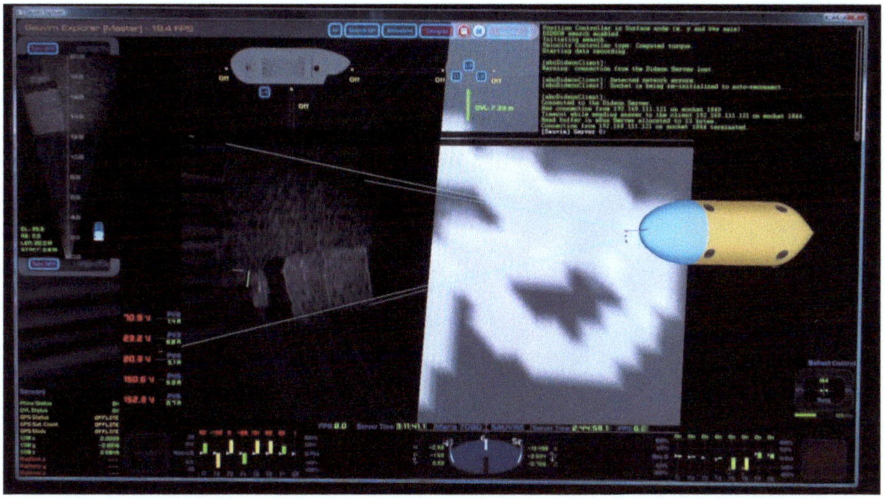

Fig. 6.17 Searching for the platform: mapping the ocean floor with the DIDSON sonar

6.2.3 Phase 3: Navigate and Dive Toward the Platform

Phase 3 includes the necessary operations to autonomously dock SAUVIM to the underwater platform.

During this phase the navigation controller uses the PHINS along with the DVL, and performs dead reckoning until the vehicle is positioned just 30 cm above the platform. The autonomous underwater docking is performed in 2 phases:

1. Surface navigation to bring the vehicle vertically over the platform
2. Vertical dive over the platform

During the vertical dive the vehicle dynamic controller operates in full 6 DOF mode, enabling the center of buoyancy identification when beyond a depth of 1 m., and with the pitch and roll optimization algorithm presented in Sect. 4.4.2.

The phase 3 is considered concluded once the DVL measures a stable distance from the platform (30 cm) for a sufficient period of time. Figure 6.18 is a snapshot of a video captured by an underwater camera during hovering. View from four cameras on SAUVIM are also shown in Fig. 6.19. Note, in particular, the upper-left view, taken from the bottom-facing camera: the vertical and horizontal reference bars mounted on the platform serve as an approximate evaluation of the position error at the end of the dead-reckoning navigation from the surface to the hovering configuration. The full navigation experiment (from the docking pier to the hovering configuration) was successfully repeated for many times.

Fig. 6.18 Autonomous docking to the underwater platftorm: view from the mini-ROV camera

Fig. 6.19 Autonomous docking to the underwater platftorm: view from the four SAUVIM cameras

Fig. 6.20 The robotic arm docked in the SAUVIM frontal bay

6.2.4 Phase 4: Hover (Station Keeping)

Phase 3 terminates with SAUVIM in hovering configuration 30 cm above the platform. This configuration is maintained for some time, which is necessary to prepare the manipulator for the next phase.

During navigation, the arm is safely docked into the front bay as shown in Fig. 6.20. The preparation consists of the following steps:

- Undock the arm
- Take the recovery tool stowed in the tool area
- Reach a forward configuration
- Activate the short-range vision system

The arm is now extracted as shown in Fig. 6.21 and ready to begin the short-range search.

6.2.5 Phase 5: Hook a Recovery Tool to the Target Object (Autonomous Manipulation)

After reaching the forward configuration with the recovery tool secured by the end-effector, the manipulator starts searching for the target object by using its optical camera, running the algorithms described in Sect. 5.3.1. Once the target object and its location (6 DOF) are detected, the arm enters a tracking mode, visual-servoing at a fixed relative position w.r.t. the target. The visual-servo target position, shown in Fig. 6.22, is computed by placing the hook at about 5 cm over the target rod, centered between the two spheres.

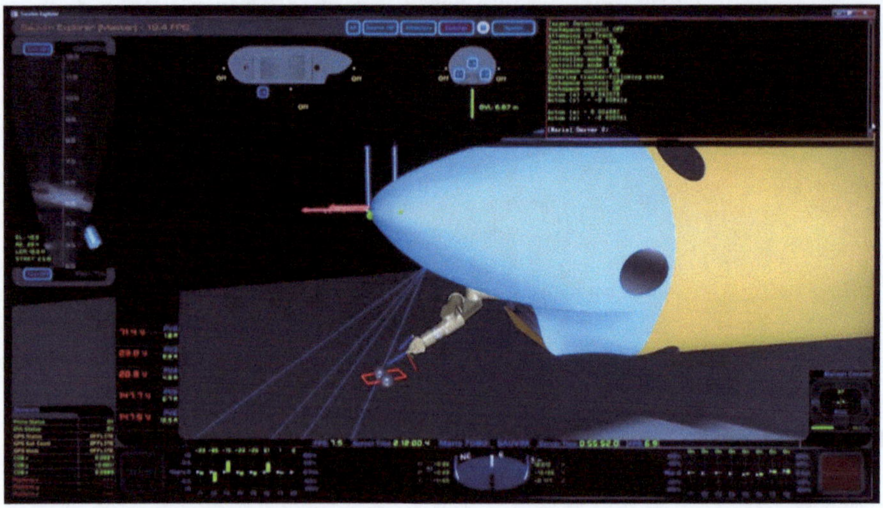

Fig. 6.21 Autonomous manipulation with workspace optimization: real time view from Sauvim Explorer

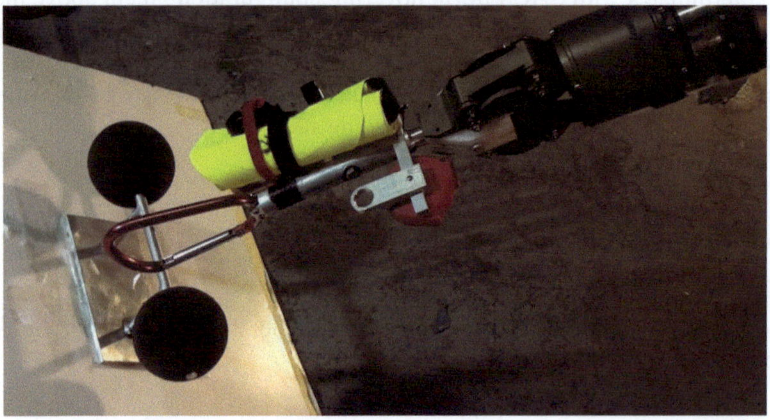

Fig. 6.22 The recovery device

The action of connecting the carabiner is triggered by simply evaluating in real-time the visual servoing error: once the error is confined within a predefined limit for a sufficient amount of time (about 5 s), MARIS physically connects the recovery tool to the target object. Figure 6.22 shows the recovery device during one of our dry tests. It consists of an inflatable buoyancy bag connected to a CO_2 cartridge. The gas flow is automatically triggered by the action of pulling a cable, rolled into the red spool, and connected to the handle.

One important feature during this phase is the workspace optimization, described in Sect. 4.2. As noted there, the vehicle is set to adjust its position in real-time for optimizing the manipulation. It is worthwhile to note that errors of the dead-reckoning navigation sensor (PHINS+DVL) were sometimes too large due to various reasons (for example, fishes swimming under the transducer) and corrections were needed. The translational difference of the two frames shown in Fig. 6.21 is an example of a typical correction.

6.2.6 Phase 6: Return to the Pier

After accomplishing the mission, SAUVIM is set to return to the docking pier.

Similarly to the initial phase, docking is performed mainly via surface navigation, but with an important enhancement: feature-based position correction. In fact, the accuracy of the DGPS would be insufficient for precise docking to the initial location (see Fig. 6.16). Precise docking is helped by "measuring" the position of a known underwater object close to the docking area, and then performing dead reckoning to the docked configuration.

For simplicity, we used the submerged platform as a known underwater object. Steps to perform the re-entry are hence the following:

- *Surfacing*. This operation is a simple vertical navigation from the hovering position to the surface.
- *Navigate to the platform search area*. Similarly to the initial phase, the navigation to the center of the harbor is a 3 DOF surface navigation, with roll, pitch and depth control inactive. The DGPS delivers the necessary measurement updates to the PHINS, with an accuracy error of about 2 m.
- *Search for the platform*. The vehicle performs again a circular scan of the ocean bottom at the maximum sonar range (40 m), keeping the platform pose identification algorithm active. The scan terminates once the platform has been detected by the pose estimation algorithm. From then, the vehicle enters a tracking mode by performing position adjustments until it reaches the predefined relative position with the platform.
- *Dead-reckoning navigation to the docking area*. By acquiring the 6 DOF pose of the platform, SAUVIM is now able to compute the relative path to the docking pier and engage it through dead-reckoning navigation.

The surface navigation of the last phase is performed by removing the DGPS position measurements from the PHINS, to avoid polluting the precise correction. The PHINS, however, still benefits from DVL velocity measurements. The measured standard deviation of the final position, despite the dead-reckoning navigation, is about 10 cm that is accurate enough for precise parking maneuvers and aligning the vehicle to the pier.

6.3 Conclusions

In this book we presented the full research path beyond the U.S. Navy funded underwater vehicle, SAUVIM project.

Autonomous manipulation, the key technology in underwater intervention performed with autonomous vehicles, is generating an increasing interest among not only the underwater community, but also the entire field of robotics.

SAUVIM has experimentally demonstrated autonomous manipulation. Following its success, more underwater vehicles having the capability of autonomous manipulation are expected to develop in the near future.

Across this monograph we have implicitly identified the complexity of problems encountered in autonomous underwater interventions. Among those, the following are most noticeable:

- **Target area navigation.** One of the most important aspects in autonomy is the environment perception. Without the help of advanced underwater imaging technology, it would have been impossible to autonomously locate and navigate toward the target area. Our presented approach is one of the first attempt to use the DIDSON sonar as multi-DOF pose sensor to guide autonomous navigation. To ensure a high success rate, it is indispensable to enhance the sensing technology and the capability of precisely assessing position measurements of various targets.
- **Vehicle positioning.** For ROVs it is the responsibility of the operator to position the vehicle in the target area. However, for AUVs the vehicle must be capable of performing workspace optimization by itself. For SAUVIM, a large effort has been devoted to develop automatic optimization algorithms capable (as seen in phase 5) of precisely correcting the hovering position for maximizing the manipulability of the arm during its operations. This important aspect is not only encountered in underwater intervention, but also in any instance of autonomous mobile manipulation.
- **Arm control system.** Many problems commonly seen in robotics such as kinematic singularities, collisions, and joint limit and motor saturation, are no longer handled by the robot operator. An autonomous robotic control system must be able to intelligently assess all the above situations and successfully carry out the given mission when it is, at least theoretically, possible. This is another area to which SAUVIM research has contributed.

Future development in autonomous manipulation should be able to provide robust improvements to all the issues above. This is important in increasing the effectiveness of an autonomous system in general and hence in increasing the confidence of all the organizations toward autonomous intervention vehicles.

NASA is one of the first organizations who have already initiated active research toward autonomous robotics in space. The Satellite Servicing Capabilities Office (SSCO[2]) is advancing the state of robotic servicing technology for regularly servicing

[2] http://ssco.gsfc.nasa.gov

satellites that were not designed with servicing in mind. Many similarities exist between underwater robotic intervention and space servicing: in fact, the research on motion control software at SSCO is based on the work presented in Chap. 3 (see Pellegrino and Roberts [20]).

In conclusion, it is our hope that this book including the findings and implications of SAUVIM would inspire the robot research community to further investigate critical issues in autonomous manipulation and to develop robot systems that can profoundly impact our society for the better.

References

1. Benjamin MR (2007) Software architecture and strategic plans for undersea cooperative cueing and intervention. In: Kaastra J (ed) White paper. National Technical Information Service
2. Newman P (2003) Moos—a mission oriented operating suite. Technical Report OE2003–07, MIT Department of Ocean Engineering
3. Albus J, McCain H, Lumia R, of Standards USNB (1987) NASA/NBS standard reference model for telerobot control system architecture (NASREM). NBS technical note, U.S. Department of Commerce, National Bureau of Standards. http://books.google.com/books?id=9aPiSgAACAAJ
4. Puts P, Elfving A (1992) ESA's control development methodology for space A&R systems. In: Jamshidi M et al. (eds) Robotics and manufacturing: recent trends in research, education, and application 4:487–492
5. MacDonald B, Yuen D, Wong S, Woo E, Gronlund R, Collett T, Trépanier FE, Biggs G (2003) Robot programming environments. In: ENZCon2003 10th electronics New Zealand conference, University of Waikato, Hamilton
6. Biggs G, MacDonald B (2003) A survey of robot programming systems. In: Proceedings of the Australasian conference on robotics and automation, CSIRO, Brisbane
7. Pembeci I, Hager GD (2002) A comparative review of robot programming languages. Technical report, The Johns Hopkins University. http://www.cs.jhu.edu/CIRL/publications/pdf/pembeci-review.pdf CIRL Lab Technical Report
8. Dai X, Hager G (2002) Specifying behavior in C++. In: International conference on robotics and automation, vol 1, pp 153–160
9. Hardin D, Frerking M, Wiley P, Bollella G (2002) Getting down and dirty: device-level programming using the real-time specification for java. In: Symposium on object-oriented real-time distributed computing, pp 457–464
10. Kanayama Y, Wu C (2000) Its time to make mobile robots programmable. In: Proceedings of IEEE international conference on robotics and automation, vol 1, pp 329–334
11. Hudak P, Fasel J (1992) A gentle introduction to Haskell. ACM SIGPLAN Not 27(5):1–164
12. Soetens P (2013) RTT: real-time toolkit. http://www.orocos.org/rtt. Accessed Nov 17 2013
13. Bruyninckx H, Soetens P, Koninckx B (2003) The real-time motion control core of the Orocos project. In: IEEE international conference on robotics and automation, pp 2766–2771
14. Soetens P, Bruyninckx H (2005) Realtime hybrid task-based control for robots and machine tools. In: IEEE international conference on robotics and automation, pp 260–265
15. Soetens P (2006) A software framework for real-time and distributed robot and machine control. PhD thesis, Department of Mechanical Engineering, Katholieke Universiteit Leuven, Belgium. http://www.mech.kuleuven.be/dept/resources/docs/soetens.pdf
16. Robot Operating System Wiki (2013) ROS: robot operating system. http://wiki.ros.org. Accessed Nov 17 2013
17. The Player Project (2013) The player project. http://playerstage.sourceforge.net. Accessed Nov 26 2013

18. Lesk M, Schmidt E (1986) LEX—a lexical analyzer generator. Bell Laboratories, unix programmer's supplementary documents, vol 1 (ps1) edn
19. Johnson SC (1979) Yacc: yet another compiler compiler. In: UNIX programmer's manual, vol 2, Holt, Rinehart, and Winston, pp 353–387, AT&T Bell Laboratories Technical Report
20. Pellegrino JF, Roberts BJ (2013) Robotic servicing technology development, American Institute of Aeronautics and Astronautics, chap robotic servicing technology development. SPACE Conferences & Exposition. doi:10.2514/6.2013-5339, http://dx.doi.org/10.2514/6.2013-5339,0

Appendix A
Mathematical Supplement

A.1 Versor Lemma

Given two frames $\langle a \rangle$ and $\langle b \rangle$ (Fig. A.1), with the latter rotated with respect to the former of an angle θ around the versor $\boldsymbol{v}$, the following relations hold:

$$(\boldsymbol{i}_a \times \boldsymbol{i}_b) + \left(\boldsymbol{j}_a \times \boldsymbol{j}_b\right) + (\boldsymbol{k}_a \times \boldsymbol{k}_b) = 2\boldsymbol{v}\sin(\theta) \tag{A.1}$$

$$(\boldsymbol{i}_a \cdot \boldsymbol{i}_b) + \left(\boldsymbol{j}_a \cdot \boldsymbol{j}_b\right) + (\boldsymbol{k}_a \cdot \boldsymbol{k}_b) = 1 + 2\cos(\theta). \tag{A.2}$$

A.1.1 Algorithm

It is possible to compute $\boldsymbol{v}$ and $\theta \in (-\pi, \pi)$ in the following way:

1. Compute first the left-hand side of Eq. (A.2) and solve Eq. (A.2) for $\delta = \cos(\theta)$
2. If $\delta = 1$ then we can assume
$$(\boldsymbol{v}, \theta) = (\forall, 0) \tag{A.3}$$
3. If $|\delta| < 1$ compute the left-hand side of Eq. (A.1) and solve Eq. (A.1) for $\boldsymbol{\sigma} \stackrel{\Delta}{=} 2\boldsymbol{v}\sin(\theta)$; our solution is hence:
$$(\boldsymbol{v}, \theta) = \pm\left[\frac{\boldsymbol{\sigma}}{|\boldsymbol{\sigma}|};\ \tan^{-1}\left(\frac{|\boldsymbol{\sigma}|}{2}, \delta\right)\right] \tag{A.4}$$
4. If $\delta = -1$ we have:
$$(\boldsymbol{v}, \theta) = \pm(\boldsymbol{v}_0, \pm\pi) \tag{A.5}$$

G. Marani and J. Yuh, *Introduction to Autonomous Manipulation*,
Springer Tracts in Advanced Robotics 102, DOI: 10.1007/978-3-642-54613-6,

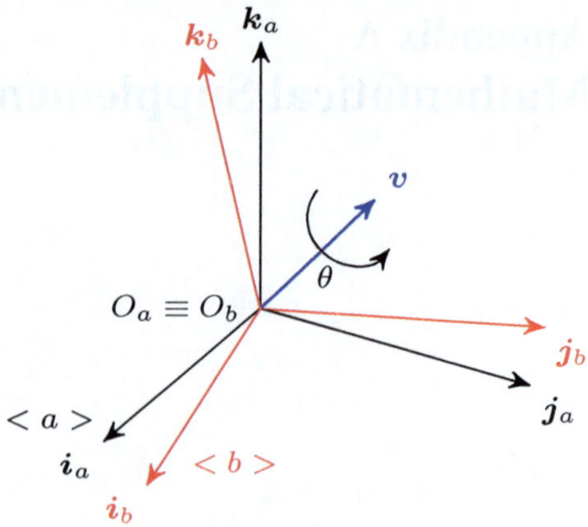

Fig. A.1 Versor Lemma: frame $\langle b \rangle$ is obtained by rotating frame $\langle a \rangle$ of $15\,^\circ$ around vector $\boldsymbol{v} = [\,0.54\ 0.54\ 0.65]^T$

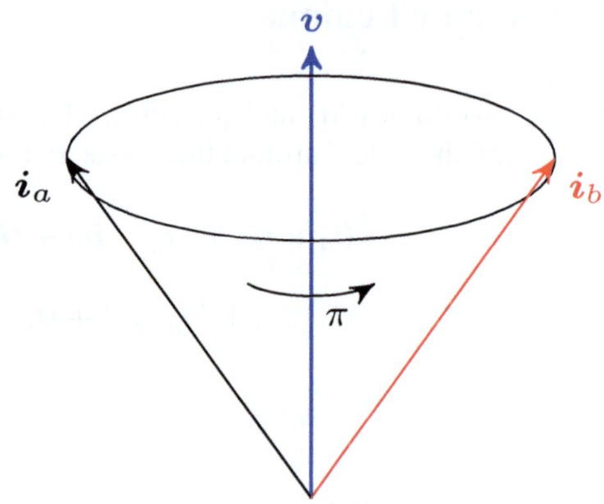

Fig. A.2 Versor Lemma: case of $\theta = \pm\pi$

where:

$$\boldsymbol{v}_0 \stackrel{\Delta}{=} \frac{(\boldsymbol{i}_a + \boldsymbol{i}_b) + (\boldsymbol{j}_a + \boldsymbol{j}_b) + (\boldsymbol{k}_a + \boldsymbol{k}_b)}{\left|(\boldsymbol{i}_a + \boldsymbol{i}_b) + (\boldsymbol{j}_a + \boldsymbol{j}_b) + (\boldsymbol{k}_a + \boldsymbol{k}_b)\right|}. \tag{A.6}$$

A.1.2 Proof of Eq. (A.6)

The algorithm is easy to understand, except for Eq. (A.6) that needs a few more considerations.

In case of $\delta = -1$, obviously the rotation angle can be only $\theta = \pm\pi$, and the vector $\boldsymbol{\sigma} \stackrel{\Delta}{=} 2\boldsymbol{v} \sin(\theta)$ is zero. Thus it is impossible to use Eq. (A.4).

However, we can note that, when $\theta = \pm\pi$, every pair of corresponding frame axis (of the two frames) is necessarily placed over a diameter of a circle bisected by the axis $\boldsymbol{v}$ (see Fig. A.2).

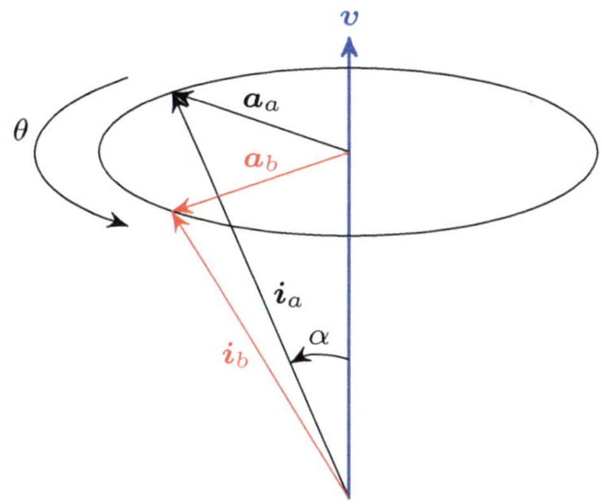

Fig. A.3 Versor Lemma: general case

Equation (A.5) immediately follows for geometric construction.

In Eq. (A.6) we have chosen to add and normalize all the vectors. This is more efficient of using the sum of only a pair of axis, which can be zero in case that the axis of rotation $\boldsymbol{v}$ coincides with one of the axis of the frame.

A.1.3 Proof of Eq. (A.1)

Consider the revolution of the versor $\boldsymbol{i}_b$ for moving from its initial position (coincident with $\boldsymbol{i}_a$) to the final (Fig. A.3). We have:

$$\boldsymbol{a}_a \times \boldsymbol{a}_b = \boldsymbol{v} \sin(\alpha) \sin(\alpha) \sin(\theta) \tag{A.7}$$

Similarly:

$$\boldsymbol{b}_a \times \boldsymbol{b}_b = \boldsymbol{v} \sin(\beta) \sin(\beta) \sin(\theta) \tag{A.8}$$

$$\boldsymbol{c}_a \times \boldsymbol{c}_b = \boldsymbol{v} \sin(\gamma) \sin(\gamma) \sin(\theta) \tag{A.9}$$

$$\begin{aligned} &\boldsymbol{a}_a \times \boldsymbol{a}_b + \boldsymbol{b}_a \times \boldsymbol{b}_b + \boldsymbol{c}_a \times \boldsymbol{c}_b \\ &\quad = \boldsymbol{v} \sin(\theta) \left(\sin^2(\alpha) + \sin^2(\beta) + \sin^2(\gamma)\right) \\ &\quad = \boldsymbol{v} \sin(\theta) \left(3 - \cos^2(\alpha) + \cos^2(\beta) + \cos^2(\gamma)\right) \\ &\quad = 2\boldsymbol{v} \sin(\theta) \end{aligned} \tag{A.10}$$

since $\cos^2(\alpha) + \cos^2(\beta) + \cos^2(\gamma) = 1$, because it is the modulus of $\boldsymbol{v}$. In fact, the quantity $|\boldsymbol{v}| \cos(\alpha) = \cos(\alpha)$ is the projection of $\boldsymbol{v}$ onto $\boldsymbol{i}_a$, and so on.

Index

G. Marani and J. Yuh, *Introduction to Autonomous Manipulation*,
Springer Tracts in Advanced Robotics 102, DOI: 10.1007/978-3-642-54613-6,